CHEMICAL WAVE TRANSMISSION
IN NERVE

Chemical Wave Transmission in Nerve

by

A. V. HILL, F.R.S.

Foulerton Research Professor of the Royal Society
Honorary Fellow of King's College, Cambridge

BASED ON THE LIVERSIDGE LECTURE DELIVERED
AT CAMBRIDGE ON 13 MAY 1932

CAMBRIDGE

AT THE UNIVERSITY PRESS

1932

CAMBRIDGE UNIVERSITY PRESS
Cambridge, New York, Melbourne, Madrid, Cape Town, Singapore, São Paulo, Delhi

Cambridge University Press
The Edinburgh Building, Cambridge CB2 8RU, UK

Published in the United States of America by Cambridge University Press, New York

www.cambridge.org
Information on this title: www.cambridge.org/9780521115445

First published 1932
This digitally printed version 2009

A catalogue record for this publication is available from the British Library

ISBN 978-0-521-05260-3 hardback
ISBN 978-0-521-11544-5 paperback

CONTENTS

PREFACE

IN his will the late Professor Liversidge made a bequest to the Master and Fellows of Christ's College, Cambridge, "for the encouragement of research in Chemistry not in ignorance of the fact that there are already in existence other Lectureships in Chemistry but because there are none such as I contemplate namely for the express encouragement of research and for the purpose of drawing attention to the research work which should be undertaken and because having regard to the vastness of the subject I wish the subject to be elucidated by as many workers as possible and feel that the friendly emulation of the lecturers holding the various lectureships above mentioned may be of benefit".

He enjoined that "if possible the lectures shall be published...so as to disseminate the information for the benefit of such of the public as are unable to attend", and further that "the lectures shall be upon recent researches and discoveries and the most important part of the Lecturer's duty shall be to point out in what direction further researches are necessary and how he thinks they can best be carried out".

"Having regard to the vastness of the subject" the Master and Fellows invited, to give the lecture in 1932, not a Chemist but a Physiologist: who, gladly accepting the challenge and its implication that physiology is a branch of chemistry (as of several other sciences), chose for his subject one in which the help of Chemists (not to mention Physicists and Engineers) is an essential condition of advance. The most important part of his duty being to point the way to further researches and "how he thinks they can best be carried out", he has urged that Physiologists should seek the aid of Chemists in the study of one of the most fundamental of their problems.

For the Lecturer has observed that Chemists (not to mention Physicists and Engineers) are very intelligent people. Often, to be sure, they are singularly ignorant, not seldom they are quite unaware of the most elementary facts. He has met—incredible it may seem—persons of great distinction in these subjects who did not know that a frog's heart will go on beating, its nerves transmitting messages, its muscles contracting, long after these are removed from their owner. He has admired the astonishment with which they regard quite simple everyday things relating to life, their readiness to accept vital phenomena as magic beyond reason or experiment. He sympathizes with them in this, for he too is often afflicted with the same wonder and astonishment. But perhaps that is because, as Ernest Starling once told him, he too is so ignorant.

One of the Lecturer's fond dreams is of a day when all educated people will have at least an elementary knowledge of the main facts of life. At present they are apt to take no interest at all in the matter: or—what is worse—to imagine that life is merely rather complicated colloid chemistry (it certainly *is* rather complicated). For biology, after all, is the fundamental science: indeed the behaviour of the nervous system—the study of which belongs to physiology—is the ultimate basis of all education and of all intellectual activity. Our bodily habits affect even our theories of the nature of things, as witness the influence of ball games on doctrines of the constitution of matter proposed by British Physicists. Our conceptions of time and space and mass depend on the impressions which reach us from our sensory organs. No Biologist has not read of the atomic theory of matter: yet many Chemists—one might say most Chemists—have never even heard of the "atomic theory" of nervous activity, namely, that this depends on wave-like impulses of an "all-or-none" character, with properties as clear and as peculiar as those of any other wave. It is sad but true, and something must be done about it.

Realizing then, on the one hand, how little they know of such things, and on the other how much interest they would find if they knew and how much help they could give, the Lecturer has broadcast this, his S O S; in the hope that a Chemist or two may be induced thereby to come to the aid of Physiologists in one of the most difficult—and therefore the most attractive—of all scientific problems, the nature of the change (he forbears to call it the mysterious change!) which is transmitted in nerve.

A. V. H.

UNIVERSITY COLLEGE, LONDON

July 1932

CHEMICAL WAVE TRANSMISSION IN NERVE

THE PROBLEM

I AM glad indeed to have this opportunity of presenting to an audience of chemists (and I hope there may be physicists and engineers among them) a problem in physiology which is one of the most fundamental of its kind and seems likely to need all the resources of science for its solution: and I am very grateful to the Master and Fellows of Christ's College for their invitation to give the Liversidge Lecture which provides the opportunity. I have borne in mind that my duty is to speak to chemists: if there be physiologists present they come at their own risk, the risk, namely, of hearing much that they know already. The opportunity I welcome is to discuss one of the essential problems of biology, stripped for once, so far as may be, of its biological clothing, exposed in its full chemical and physical nakedness.

The title chosen assumes that the "something" which is transmitted in nerve, when an impulse runs along it, has some of the properties of a wave, and that these properties can best be described in chemical, or strictly speaking in physico-chemical, terms. It does not imply that chemistry is able now, or will be able some day, to explain all the behaviour of the nervous system: no such hypothesis is necessary, as Laplace said of another matter: indeed, for all I know, the boot may prove to be on the other leg: it may some day be necessary to invoke the properties of the human nervous system to explain the theories which chemists, or at any rate physicists, invent to account for the phenomena they observe! It may be that the discontinuous nature of sensation and, so far as we are aware, of all nervous action, including probably thought itself, the fact that these depend on the transmission not of continuous states but of discrete and separable wave-like impulses, allow only certain types of phenomena and relations to be perceived and appreciated: so that atomic and quantum hypotheses may ultimately be found to depend upon the nervous apparatus through which the phenomena, on which they rest, have filtered. I say this, not as an expression of faith, not even to discipline my physical friends, but to show that I am not too confident a mechanist, that I am perfectly well aware that stripping a problem of its biological clothing may be like depriving a man not of his shirt but of his skin!

All our sensations, all our movements, probably most of the activities of our nervous systems, depend upon the properties of a certain transmitted disturbance which we call the nervous impulse: this to neurology is what the atom is to chemistry, the electron and the quantum to physics. Such transmitted disturbances, as a matter of fact, are not peculiar to nerve: they occur rather generally in the protoplasm of cells, not only of animals but of plants. A rapid reaction to events occurring at a distant point may be necessary for efficient working or for safety. Most living cells, however, are small, so that the distances involved are usually a few thousandths, or at most a few hundredths, of a millimetre: and over such small distances the velocity of transmission being of no great concern, a highly differentiated process is unnecessary. As soon, however, as larger complex animals were evolved, the problem of reaction to events occurring at a distance became more acute, and special cells were developed to deal with it. These cells are still small in all dimensions but one: a fine thread, however, the axon or nerve fibre, runs out from them, which is only $3-25\mu$ (0·003–0·025 mm.) in diameter, but in the largest animals may be many metres in length. Along these threads messages—not material substances—are sent. We are concerned with the nature of these messages.

THE VELOCITY OF NERVE IMPULSES

Medullated nerve, mammal,[1] 37° C., about 100 m./sec.
Medullated nerve, dogfish,[2] 20° C., about 35 m./sec.
Medullated nerve, frog,[1] 20° C., about 30 m./sec.
Non-medullated nerve, crab,[5] 22° C., 5 and 1·5 m./sec.
Non-medullated nerve, mammal,[3] 37° C., about 1 m./sec.
Non-medullated nerve, olfactory of pike,[4] 20° C., 0·2 m./sec.
Non-medullated nerve, in fishing filament of *Physalia*,[3] 26° C., average 0·12 m./sec.
Non-medullated nerve, in Anadon,[1] 0·05 m./sec.
Compare the velocity of sound in air at 0° C., 331 m./sec.

[1] See Broemser (1929). [2] See Monnier and Monnier (1930, p. 14).
[3] See Parker (1932 *a*). [4] See Nicolai (1901).
[5] See Monnier and Dubuisson (1931).

THE NERVOUS SYSTEM

Each nerve fibre depends, for its continued existence, on an intact connection with its cell—it is part of the cell (see fig. 1). Cut it at any point in its course and the peripheral portion (that remote from the cell) loses its function and dies. The " degeneration " takes a few days and travels progressively down the fibres. It is not the absence of normal activity which causes loss of function,

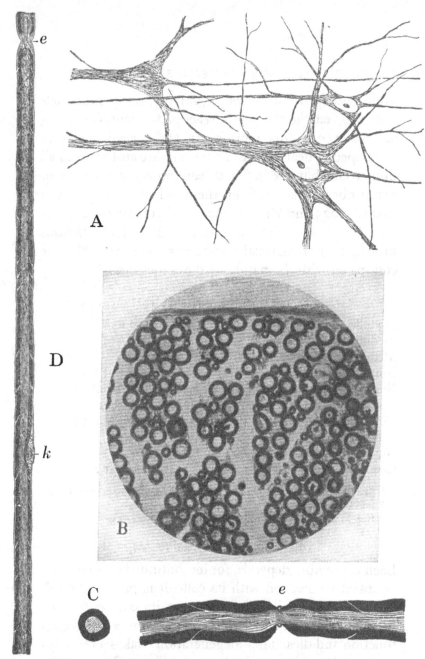

Fig. 1. *A*. Three nerve cells with "processes". The nerve fibres shown in *B*, *C* and *D* arise as processes from such cells: other processes maintain functional connection between the cells. *B*. Section of the sciatic nerve of a cat, showing the variation in size of its constituent fibres. The black rings are the sheaths, the white areas inside are the axis cylinders. *C*. Diagram of medullated nerve fibre on larger scale, transverse and longitudinal sections. *e*, node of Ranvier. *D*. Medullated nerve fibre. *e*, node of Ranvier; *k*, nucleus. In *A* the preparation was treated with Ramón y Cajal's silver nitrate "photographic" method: in *B*, *C* and *D* with osmic acid, which stains the lipins black.

for in a sensory nerve such normal activity starts from the peripheral end where it can still start after the lesion. It must be the absence of a connection with the nerve cell which makes the axon degenerate. The latter will grow down again from above if the cut ends are kept in juxtaposition, but that is a slow matter.

At first it might have seemed that the cell is supplying something material which is passed along the intact fibre. The distances involved, however, may be so great that some very special method of passage would be required: as Sir W. B. Hardy once said, it would require geological time for diffusion to work between the spinal cord and the tail of a whale! (see also Hardy, 1927). Moreover, with a surface so relatively large, 2000 sq. cm. or so per gram, and a wall so thin, there could scarcely fail to be loss on the way. I doubt therefore whether any material substance travels far along the fibre, although Parker (1932 b) thinks that some hormone-like controlling substance percolates down the nerve from the nucleus of the cell. I am more inclined to believe that its state is maintained by an influence of some sort, by the field of one molecule on the next. If so, here is another type of transmitted effect, one which physical chemistry may help to make intelligible; it is altogether different, however, from the rapid wave-like impulses which make up nervous activity, and we will return to these.

In a telephone system it is impossible to arrange that each subscriber shall be connected with every other by a separate line: exchanges and groups of exchanges are required. Similarly in animals where it is necessary that any part shall be able, on occasion, to call for reaction in any other part, ganglia and groups of ganglia are needed: a central nervous system comes into being, based in its highest forms largely upon the distance receptors, particularly on those which are sensitive to light. Into this central organ most of the cells themselves (*A*, fig. 1), as distinguished from the fibres in which the messages go (*B, C, D*), are collected. This is not the occasion to discuss the central organ itself: those who wish to know more about it and the way in which messages are sorted and co-ordinated there, are referred to a recent work, *Reflex Activity of the Spinal Cord*, by Sir Charles Sherrington and four of his collaborators (Creed, Denny-Brown, Eccles, Liddell and Sherrington, 1932).

MESSAGES—EVENTS, NOT MATTER OR ENERGY

I have spoken of messages, not material substances. It is inconceivable indeed that discrete packets of any actual substance could travel 100 m. per second and 500 times per second in either direction along a jelly-like thread only 0·01 mm. in diameter and containing 80 p.c. of water: and yet apparently not accumulate at the end, or leak over into similar threads lying within a few thousandths of a millimetre. There are no tubes in nerves through which "vital spirits", or Descartes' subtle vapour, or Borelli's "succus nerveus" might be squirted (see Borelli, 1710, Part II, p. 37): and there is no conceivable mechanism by which separate material projectiles, with the properties which I will describe, could be fired along such an unpromising channel.

Nor, on the other hand, is there any evidence that the nerve impulse is a special form of energy. We hear the expression "nervous energy", and some psychologists speak as though this peculiar entity could be "drained" along tracks in the nervous system. The expression and the mental picture which it calls up have no more scientific basis than the "iron nerves", or the "steel sinews", or the "icy stare", of common speech. They may be justified when we speak in parables, but they have no place in scientific discussion. Energy, as we shall see, is liberated during nervous activity as in all natural processes, but there is no reason to think that this energy has any peculiar properties: it is derived from chemical reaction. It is safer to avoid expressions which can be, and often are, misunderstood.

THE NERVE FIBRE

The nerve impulse then is an event, not a substance or a form of energy, and it is transmitted along a tiny thread of protoplasm which in some cases, but not all, consists of two separate parts, an axis cylinder and a sheath (see fig. 1). It then looks like an electric cable with conducting core and insulating cover. The sheath contains a large proportion of fatty substance, as is shown by its staining properties: and it has little breaks at intervals, the nodes of Ranvier, to which no definite function has yet been allotted. Its material is of high specific resistance (Appendix 1). The axis cylinder, or core, is a soft transparent thread of protoplasm, a jelly-like substance much the same as is found inside other living cells: its specific resistance is low. In non-medullated fibres, which occur in the peripheral parts of the involuntary nervous systems of vertebrates and in the finer connections of their central nervous systems, and are the only type to be found in most lower animals, nothing but the jelly-like core can be seen: the sheath

is so thin that it amounts to no more than a film, probably only a few molecules thick, such as exists at the
surface of many living cells. The structure, alas, particularly after fixing and staining, tells us little about the
function: we have to employ other methods than those
of the microscope.

The osmotic pressure of the core of a nerve is the
same as that of blood and other organs: in frogs it is that
of a 0·7 p.c. NaCl solution, in mammals 0·95 p.c., in sea
crustaceans (crabs, etc.) more than 3 p.c. This is made
up almost entirely by common ions, Na·, K·, Cl′, phosphate. In blood, of the cations Na· is in great excess:
in nerve, as in muscle, there is a large preponderance of
K·. The potassium is of great importance, as we shall
see, in relation to the electrical potential difference
normally existing across the boundary.

THE ELECTRIC "ACTION CURRENT" OF NERVE

One naturally asks—how can the presence of an impulse
in a nerve, or its arrival at the other end, be detected?
Firstly, of course, by the physiological effect—sensation
or response as the case may be. That, however, requires
an intact animal, or at least that the nerve should be
connected to some organ, e.g. a muscle, to effect the
response. It is fortunate, therefore, that isolated nerves
work well for hours, indeed for days, after separation
from their original owner, and that another method is
available for detecting the presence of an impulse in a
nerve—the method of recording the electric change
which goes with it. This electric change seems to be a
universal accompaniment of the impulse, it can be used
as a sign of its presence, as a measure of its size.

No difference of electrical potential can be detected in an uninjured resting nerve: without injury or stimulation we can examine only the uniform outside surface. If, however, a nerve be injured, as by cutting it, a potential difference is found, of the order of a few hundredths of a volt, between the injured and the uninjured points, in the sense that positive current runs in an external circuit towards the injured part. The source of this potential

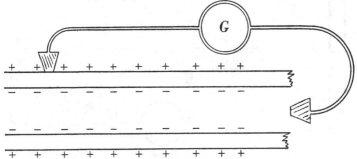

Fig. 2. The polarized state at the surface of a nerve fibre, shown by means of an injury at one end and two electrodes placed, one at the injured and the other at an uninjured point. By using a potentiometer, instead of the galvanometer G shown, the potential difference may be measured. Allowing for the degree to which it is short-circuited by the conducting fluids around the fibre it is probably of the order of 50 mv. In crab's nerve, with a strong salt solution between the fibres, it is regularly measured as 30 mv.

difference is probably a polarized state (fig. 2) in the surface membrane of the nerve fibre, the electrode at the injured point making contact, more or less effectively, with the inside of the fibre, the inner surface of the membrane. The injury *does not produce* the potential difference, it merely allows it to be manifested. In certain very large plant cells direct contact can be made with the inside of the cell by a capillary electrode and analogous electrical phenomena can be shown (see Osterhout, 1931, p. 381, etc.).

If two electrodes be placed upon a nerve and the nerve

be stimulated, e.g. with a single rapid condenser discharge or a single induction shock, a momentary change travels along it which can be recorded with the cathode ray oscillograph, or some other suitable instrument. At any given instant a certain length of the nerve, of the order of a few centimetres, is found to be the site of a

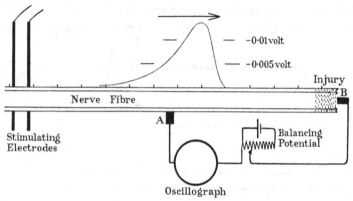

Fig. 3. The action current in a medullated nerve fibre at about 6° C. Distances marked in cm. A shock is given at the electrodes shown on the left, and the impulse travels to the right at about 10 m. per second. The action current ("monophasic") is picked up by electrodes at an uninjured point A and an injured point B, so as to avoid the complication of a diphasic record, which occurs if the wave passes both electrodes. The "injury" or resting potential between A and B is balanced and the potential difference caused by the passage of the wave is recorded on the oscillograph. It follows the course represented by the dotted line, occupying at any given instant a few centimetres length of the nerve.

wave of negative potential, that is to say a positive current will run, in an external circuit, from a resting to an active point. The amplitude of this wave is a few hundredths of a volt: it moves at a speed depending on the nerve and the temperature—anything from 0·05 to 100 m. per second. We are probably right in thinking that the impulse itself, whatever it is, occupies the same region, and moves at the same speed, as its electrical accompaniment (see figs. 3, 4, 5, 6).

THE ELECTRIC ORGAN

A nerve is composed of many parallel fibres, and if the wave of activity had the same velocity in each, the individual "action potentials" would remain in parallel and the potential changes of the whole wave, apart from short-circuiting by the inactive fluid between them, would be identical with those of the individual fibres. Actually there is considerable short-circuiting and the velocities are different, so the whole wave gradually spreads out (fig. 6) as it moves along and has a lower potential than the ideal potential of the single fibre. There is, however, an organ present in some animals, the electric organ, in which the elements are arranged, not in parallel but in series, and provision made for their simultaneous activation. In the Japanese electric ray, for example, the electric organ consists of 200 hexagonal prisms in parallel, each 1–1·5 cm. high, and each consisting of 400 electric plates arranged in series. Each plate is a clear jelly-like mass, a cell with large nuclei. The organ is arranged in two halves, one on each side of the body. There are five electric nerves on each side, which branch and end on the ventral side of each plate. The nerves are simultaneously excited from the electric lobe of the brain.

When excited the ventral side of each plate develops for a moment a negative action potential, just like the surface of a nerve, but the elements are here in series and summation occurs. A difference of potential of 30 volts may thus be set up between the ventral and the dorsal surfaces of the animal, the latter being positive. The impulse lasts for about 0·01 sec., a time sufficient for a very effective stimulus to another animal. A series of such discharges in rapid succession, particularly with the high electrical conductivity of sea water, can produce a stunning effect.

In embryological development the electric plates are analogous to muscle cells, which show electrical phenomena similar to those of nerve. There is no reason to think that the shock given by the electric organ is, in any fundamental sense, different from the action current of nerve. In the former the elements are in series, in the latter in parallel: the nature, however, of their activity is almost certainly the same. In nerve, taking account of short-circuiting between the fibres, the "injury" or resting potential is generally estimated as 25–50 mv.: in the electric organ, 400 elements in series, producing a total of 30 volts, must develop an average of 75 mv. apiece. The order of size is the same. For further details of the electric organ see Rosenberg (1928).

OTHER SIGNS OF ACTIVITY IN NERVE

In addition to the electric change, which is the sign most commonly used in investigation, there are three other accompaniments of activity in nerve as well as the physiological effects, sensation and response. These are, and they will be referred to in detail later, heat formation, oxygen consumption and carbon dioxide production. Their importance is that other chemical changes are difficult as yet to detect: not that they do not occur, but because they are too small. Heat production, therefore, which is a sign of chemical change, is a valuable indicator of what is happening inside the nerve. Heat, moreover, can be measured with little lag, and—what is particularly important—without injury to the living cell: nearly all chemical treatment destroys it.

PROPERTIES OF THE NERVE IMPULSE

Let us consider now the properties of the impulse which is transmitted, or transmits itself, in nerve; these properties have been discovered, partly by the use of a

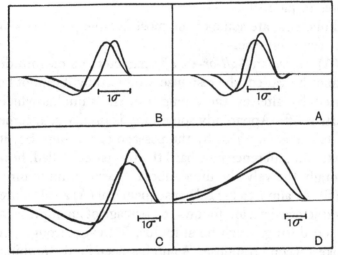

Fig. 4. Single action potential waves, in single nerve fibres, recorded by the Matthews' oscillograph. Superimposed tracings from two records, one—the more rapid—from a nerve fibre connected to a sensory end-organ in a muscle, the other—the slower—from a nerve fibre connected to a sensory end-organ in the skin. *A*, *B* and *C* are diphasic, i.e. both electrodes are on uninjured points, *D* is monophasic. See fig. 3 and legend. *A*, 14·8° C., electrodes 18 mm. apart; *B*, 15° C., electrodes 10 mm. apart, the second phase therefore coming on more rapidly; *C*, 10·5° C., electrodes 12·5 mm. apart, the lower temperature slowing the response; *D*, 10·5° C. (From Matthews, 1929, fig. 3, p. 177.)

preparation consisting of muscle and nerve, partly by means of oscillograph records of action currents in nerve alone. The single impulse in the single fibre is the basis of nerve activity. Until recently this individual impulse could not be separately examined, and deductions had to be made from the results of stimulating many fibres in parallel. Recently, through improvements in technique, especially by Adrian and Matthews, it has become possible to record the form and movement of the electric change resulting from a single impulse in a single fibre (see fig. 4). No cause, however, has been found to doubt the deductions made from experiments on numbers of fibres in parallel.

Following are some of the most notable properties of the nerve impulse.

(A) It has an "*all-or-none*" character: its magnitude cannot be varied, the response produced by it cannot be varied by altering the strength of the stimulus which starts it off. Apparently something is discharged, some container is emptied, by the passage of the impulse: it makes no difference how hard the trigger be pulled, how strongly the valve be lifted, above a certain minimum.

(B) As almost a logical consequence of (A) an absolute "refractory" period follows the passage of an impulse, a period during which no stimulus, however strong, can evoke a second response. A gun has been fired—nothing can happen till it is reloaded: a cistern has been emptied —we must wait till it is filled again: a condenser has been discharged through a neon lamp—the potential difference must be restored before the discharge can take place once more. In frogs' nerve at 20° C. the absolute refractory period is about 0·001 sec.: a rise of temperature of 10° C. diminishes it to one-third: a fall of temperature of 10° C. increases it three times (Amberson, 1930). As we shall see, the return of a nerve after discharge to its initial excitable state probably involves two factors: the restoration of its normal electrical resistance to a film through which a discharge has occurred: and the rebuilding of a potential difference across the film. One at least of these factors must have the high temperature coefficient characteristic of a chemical reaction.

(C) As a direct consequence of (B) two nerve impulses going in opposite directions in the same fibre come each into the refractory region of the other and both are wiped out. Thus, unlike telephone lines, nerves cannot be used for simultaneous transmission in both directions. Separate systems, "sensory" and "motor", are required: this is the first principle in the design of the nervous system.

(D) After the absolutely refractory stage is over a relatively refractory stage persists, during which a stronger stimulus than usual is required to start an impulse, and the impulse, even when started, is smaller than the original one. Discharge, therefore, can take place before recharge is complete.

It has recently been shown (Erlanger and Blair, 1931 *a* and *b*) that even a completely ineffective stimulus (an induction shock, or a short constant current, below the "threshold" of excitation) may leave a relatively refractory condition behind it. The immediate result of an ineffective stimulus is to make the nerve more excitable: the accumulation of ions, or the partial discharge of a capacity, produced by the first shock is not abolished at once and the second need not be so strong. Later, however, for several thousandths of a second, after the first effect has worn off, the nerve is less excitable and not more.

I shall refer to this later (see pp. 45, 46 and Appendix 1): it is a very fundamental observation, and it serves to explain—as we shall see—why sufficiently frequent stimulation (not of "high frequency" in the electrical sense) may fail to produce a response in a muscle attached to a nerve ("Wedensky inhibition").

(E) As a result of the refractory phase the frequency of transmission is limited: in frog's medullated nerve at 20° C. to about 500 impulses per second: at a lower temperature to a correspondingly lower figure. The refractory phase lengthens somewhat during continuous stimulation. In some nerves the maximum frequency of response is much less. In the actual functioning of the nervous system nearly all messages consist of trains of impulses—but the single impulse in the single fibre is the quantum on which the complex of nervous activity is based. The size of the quantum, however, varies inversely with the frequency (see fig. 9).

(F) The impulse at any instant occupies a length of a few (3–7) centimetres in a nerve fibre; which is several thousand times as great as the diameter of the fibre. This length is nearly but not quite unaltered by a change of temperature, although the velocity of transmission is greatly affected thereby (Lucas, 1909; Gasser, 1928 a, 1931). The length is approximately proportional to the diameter of the fibre (see below). There must be something significant in these facts, though at present we can only guess what they mean. I suspect we shall find that the length in question depends upon some simple physical ratio, e.g. of electrical resistance axially per unit length of the core to electrical resistance radially per unit length of the sheath. Rushton (1928) has defined an "analytical unit" of length in terms of these quantities.

If the ratio of axial to radial resistance were greater the length of the wave should be less, since the local electric current (on which probably transmission depends) would not spread out so far in front of an active region and the velocity would be less: if the ratio of axial to radial resistance were less the length of the wave should be greater. The small effect of temperature on the length would then be merely a consequence of the probable fact that temperature has nearly the same influence on both of these resistances.

According to Gasser (1928 a, p. 700) "the duration of the rising phase of the action potential is the same in all the fibres of a nerve trunk"; according to Gasser and Erlanger (1927, p. 546) "the duration of the axon action potential is the same in fibres of all velocities". They add that in mammalian nerve the rising phase can be taken as 2×10^{-4} sec. and the total

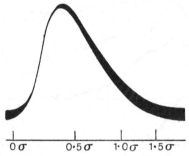

Fig. 5. Cathode-ray oscillograph record, on a negative exponential time scale, of the action potential wave in the phrenic nerve of a dog, due to a single shock. (Copied from Gasser, 1928 b.)

duration 6×10^{-4} sec.: in frog's nerve the corresponding quantities are 3×10^{-4} and 9×10^{-4} sec.

Now Gasser and Erlanger conclude (1927) that the velocity of conduction in a fibre of a given nerve trunk is determined by its diameter being approximately proportional to the diameter. The *duration* being independent of the diameter, the length occupied by the wave must be proportional to the diameter.

Fig. 5 is a cathode-ray oscillograph record (Gasser, 1928 b, p. 571) of the action potential wave in the phrenic nerve of a dog, produced in response to a single shock. The fibres are of approximately the same size and no great spreading out of the wave has appeared.

Fig. 6 is transposed from cathode-ray oscillograph records into rectangular linear co-ordinates (Erlanger and Gasser, 1924). The separate waves represent the action potential in the sciatic nerve of the bull frog, after conduction for various distances (from 12 to 82 mm.).

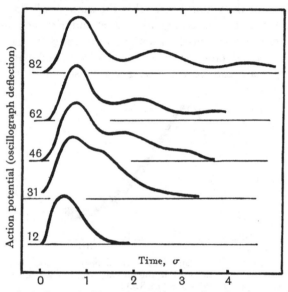

Fig. 6. Cathode-ray oscillograph records, transferred to rectangular linear co-ordinates, of the action current in a bull frog's sciatic nerve, after propagation from the site of stimulation through the distances shown on the left (mm.) Note the spreading out of the wave, owing to the lower velocity in the smaller fibres which occur in two main groups. (Copied from Erlanger and Gasser, 1924.)

The times given above for the duration of the action potential at any point are probably of the same order of size as Fr, where F is electrical capacity and r is resistance perpendicular to the axis, per unit length of nerve fibre sheath. The coincidence may well be significant (see Appendix 1).

(G) The different fibres of a given nerve trunk are of different diameters, and impulses travel in them with velocities proportional to their diameters (Gasser and

Erlanger, 1927). Here too is an obviously significant fact. I suspect that, like those just mentioned, it depends upon some physical relationship, e.g. the speed of discharge of a capacity distributed along the surface of the nerve fibre.

The capacity per unit length, depending on the surface, would increase as the first power of the diameter, the resistance through which it discharges (if chiefly that of the axis cylinder, see fig. 10) inversely as the square of the diameter. The time of discharge, being proportional to capacity × resistance, would vary therefore inversely as the diameter: in a given time a greater length would be discharged, in proportion to the diameter.

(H) When the temperature is raised or lowered by 10° C. the velocity of transmission is increased or diminished in the ratio of 1·7 to 1. If excitation from point to point is transmitted by the action current, it is natural that if this rises more rapidly its effect at a distance should be produced quicker.

(I) The passage of an impulse is associated with a liberation of heat which we will now discuss.

THE HEAT PRODUCTION OF NERVE

Muscle liberates energy when excited, heat plus work during contraction, heat about equal to the sum of these in "recovery" afterwards. In a single muscle twitch the immediate rise of temperature is about 0·003° C., and when shortening is prevented the force developed (T) bears to the heat (H) a very constant relation

$$\frac{Tl}{H} = \text{about } 5,$$

where l is the length of the muscle fibres. When mechanical work is done the heat is changed in certain characteristic ways. It is clear that the heat is directly related to the performance of work and the development of mechanical force.

By nerve no work is performed or force developed, and for long it was believed that in the transmission of nerve impulses no heat is evolved. The impulse was pictured (in Bayliss's words, 1924, p. 398) as a "reversible physico-chemical process, not associated with loss of material on account of metabolic reactions". We were all wrong. Heat is set free, but its amount till recently was too small to detect: indeed my own failure to detect it was the basis of Bayliss's opinion. In the last few years, however, it has come within the range of practical measurement and its general relations are now evident. In crab's non-medullated nerve, indeed, the stage has been reached when the heat in a single impulse could just be recorded: in medullated nerve, where the heat is far less, 100 impulses are required. Crab's nerve provides the easier material and in some ways the clearer (though not identical) results. Actually, however, because they are much more readily obtained in physiological laboratories, the properties of frogs' medullated nerves are better understood. Physiological laboratories ought to be built near the sea, or at least within reach of aquaria where delicate marine animals can be stored. The heat production of crabs' nerves could have been measured 20 years ago: actually 14 years were wasted before it was measured with frogs (see Appendix II).

The fact that heat is produced, and the manner of its production, dispose of the possibility that the nerve impulse is a "reversible physico-chemical process". This heat is our chief clue as to the intimate nature of the transmitted wave. It is the justification of the title of this lecture. We shall consider its characteristics, therefore, in some detail (see Hill, 1932).

When a single isolated impulse travels along a nerve there is an immediate heat production which we call the

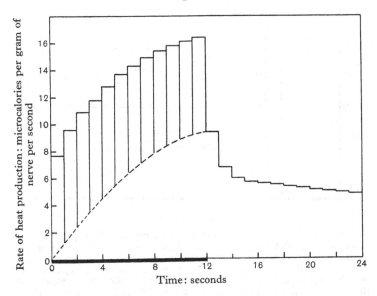

Fig. 7. Heat production of frog's nerve in oxygen, during and after a 12 seconds almost maximal stimulus at about 20° C. Horizontally, time, stimulus shown by black line beneath: vertically, rate of heat production in 1 second blocks (measured from the base line) obtained by the analysis of photographic records.

The analysis is shown only for 24 sec. The galvanometer deflection recording the heat was followed for 25 min., by which time recovery was complete, and the total heat found to be 1410×10^{-6} cal. per gram. The "initial" heat was taken as the area of the curve above the broken line, which is 90×10^{-6} cal. per gram: the rest is the recovery heat. The ratio, therefore, of total heat to initial heat is about 16.

"initial heat", occurring during, or immediately after, the actual transmission: and a later heat production, which we call the "recovery heat", diminishing rapidly in rate at first and then more slowly (in frogs but not in crabs), lasting probably for nearly an hour at 20° C. (see figs. 7, 8). We will deal later with the recovery heat—for the moment let us consider the initial heat only. The recovery heat, however, is far the more important, at least if importance depends upon size.

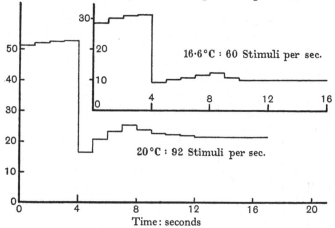

Fig. 8. Heat production of crab's nerve in oxygen, during and after a 4 seconds' stimulus. Two cases shown at 16·6 and 20° C. respectively. Vertically, rate of heat production in 1 second blocks (microcalories per gram), obtained by the analysis of photographic records (see fig. 13). Neither stimulus was quite maximal as regards frequency. Note, nevertheless, that the rate of heat production during stimulation and afterwards is much greater than in the frog (fig. 7). Note also that the characteristics of the recovery heat are different, there is no early rapid fall of recovery heat rate in the crab (Beresina and Feng).

THE "INITIAL" HEAT

At 20° C. the initial heat in a single isolated impulse in a frog's medullated nerve averages about 7×10^{-8} cal., i.e. about 3 ergs, per gram of nerve. At 0° C. it is greater, about $2 \cdot 5 \times 10^{-7}$ cal., or about 10 ergs, per gram. The ratio is about $3 \cdot 5 : 1$. The velocity of transmission is diminished about $1 \cdot 7$ times by a fall of 10° C., about 3 times by a fall of 20° C. The length of nerve occupied at any instant by the wave is little affected by a change of temperature. The initial heat, therefore, is about proportional to the *time* during which the transmitted disturbance persists at any point. During that time some reaction or change occurs—the heat is proportional to its extent. The problem is—what is the reaction or change in question?

THE ENERGY IN THE ELECTRIC CHANGE

One naturally asks—may not the electrical disturbance transmitted in nerve be the source of the initial heat? A wave of potential difference is propagated, currents flow through the nerve and the conducting solutions around it, heat must be set free. If we take the observed potential differences (fig. 3) along the nerve and assume that these cause currents to flow axially through the conducting media inside and outside the sheath, then the Joule's heat of these currents can be calculated. The result is only a small fraction, less than 1 p.c., of the observed initial heat (Hill, 1921). There is another possibility, however, which we will discuss shortly: namely, that the nerve fibre is to be regarded as a cylindrical condenser charged to 50 mv. or so, this condenser being discharged, or partly discharged, as the wave sweeps by. This possibility cannot be lightly rejected on the ground that its energy is too small: I doubt however, whether it is the real cause of the initial heat.

THE HEAT PER IMPULSE

When a nerve is stimulated not once only but many times in succession the heat per impulse is less: in fig. 9 the heat per impulse is plotted as a function of the interval between impulses. Here we see clearly some kind of recharging process at work. There is a limit to the rate at which energy can be liberated. As the frequency of stimulation is increased the heat per second (in frog's nerve unless otherwise stated) approaches a constant limiting value: at 20° C. about $7 \cdot 5 \times 10^{-6}$ cal., or about 300 ergs, per gram of nerve per second, at 0° C. about $2 \cdot 2 \times 10^{-6}$ cal., or about 100 ergs, per gram per second. At 30° C. the maximum heat per second

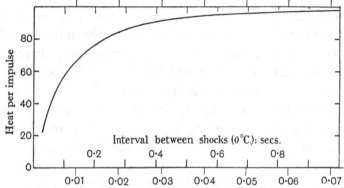

Fig. 9. Heat per impulse (arbitrary units) as a function of interval between impulses, during stimulation at 0° C. and at 21·3° C. The physical basis of this relation is discussed by Hill (1932, p. 129) and by Gerard, Hill and Zotterman (1927).

would probably be about 500 ergs per gram, and at that temperature nerve would transmit about 1000 impulses per second: the energy per impulse, therefore, at the highest frequency would be about 0·5 erg per gram. Thus, from zero frequency at 0° C. to the highest possible frequency at 30° C., the *heat per effective impulse transmitted varies from about* 10 *ergs to* 0·5 *erg per gram of nerve.* Let us try to calculate what this means.

HEAT RELATED TO SURFACE OF FIBRES

The facts (*a*) that the impulse travels with a speed approximately proportional to the diameter of the fibre, and (*b*) that the region occupied by the wave is also approximately proportional to it, strongly suggest that some surface reaction is involved. If the wave were propagated by an impulse occurring throughout the axis cylinder (like sound in the air of a pipe) there would be no obvious reason why its speed, or the region it occupies, should vary directly with diameter. If, however, we think of some reaction between the core and its surface, for example the accumulation or dissipation of an electric charge on the surface by conduction through the core, then such a relation might—and probably would—occur. We are led to think of the active region as located at a cylindrical surface running parallel to the axis. It is interesting, therefore, to calculate the energy given out during the transmission of the impulse, not per gram of nerve, but per sq. cm. of surface of the fibres of which the nerve is composed.

Let us assume, for simplicity, that medullated nerve consists of fibres 10μ in diameter, and that half its volume is filled with such fibres, the rest being connective tissue, lymph, blood vessels, etc. Let l be the total length of fibres contained in 1 gram of nerve. Then

$$5 \times 10^{-1} \text{ c.c.} = l \times \pi \times (10^{-3})^2/4 = \pi l \, 10^{-6}/4,$$
$$l = 20 \times 10^5/\pi \text{ cm.}$$

Thus we calculate that 1 gram of nerve contains about 6·4 km. length of nerve fibres.

The area of the surface* of the fibres then is

$$\pi \times 10^{-3} \times 20 \times 10^5/\pi = 2000 \text{ sq. cm.}$$

The energy, therefore, in the transmission of a single

* The surface calculated is that of the outside of the sheath: that of the axis cylinder (see fig. 1) would be about half as much.

effective impulse which, as we have shown, varies from 10 to 0·5 erg per gram of nerve, is from 5×10^{-3} to $2·5 \times 10^{-4}$ erg per sq. cm. of fibre surface: a very small quantity, *the smallness of which can be realized by the statement that it is $1/4000$ to $1/80,000$ of the surface energy of a water-olive oil interface.*

Let us imagine that some molecular change occurs over the *whole* surface by which this energy is set free. In a condensed film of fatty acid on the surface of water the area occupied by a molecule averages about 25×10^{-16} sq. cm. Assume that the surface molecules in the nerve fibres are of this order of size. The energy per molecule, therefore, would be $1·25 \times 10^{-17}$ erg for the upper, $6·25 \times 10^{-19}$ erg for the lower limit. These are very small quantities. The energy of a quantum of visible light is a multiple of 10^{-12} erg.

To get some further idea of their order of size let us calculate them *per gram-molecule* of the molecules in the surface. We multiply by 6×10^{23} and find $7·5 \times 10^6$ and $3·75 \times 10^5$ ergs respectively. Expressed in heat units these are 0·18 and 0·009 cal. respectively. It is a very slight chemical reaction indeed which gives so little total heat per gram-molecule of the reacting substance. Heats of combustion of simple organic compounds are of the order of hundreds of thousands of calories.

Even if the surface consisted of much larger molecules, e.g. of 100 times the area and 1000 times the molecular weight, the result would be scarcely less emphatic: the energy set free, therefore, in the transmission of an impulse is an extraordinarily small quantity when distributed over *all* the molecules in the surface of the nerve fibre transmitting it. It is difficult to believe that a chemical breakdown worthy of the name, involving the whole surface of the fibre, can be responsible for the propagated disturbance.

INITIAL HEAT AS ENERGY OF CONDENSER DISCHARGE

Let us take, however, quite a different basis for calculation, one to which I have already referred, and suppose that the initial heat, the energy liberated during the transmission of the impulse, is derived from an electrical source, namely the discharge of a condenser, of an electrical double layer of some kind, located at some surface in the fibre. Let us assume that the resting potential of nerve is 50 mv., and that the capacity per gram of nerve is F μF. The energy of a condenser of capacity F μF charged to voltage V is $5 FV^2$ ergs. At 0° C. if we assume the energy of the condenser to be half discharged (the fraction taken is arbitrary) the initial heat in a single impulse, which is 10 ergs per gram as we have seen, should be equal to

$$\tfrac{1}{2}\, 5F\,(0\cdot05)^2 = 6\cdot25 \times 10^{-3}\,F.$$

Hence $F = 1600$ μF, which is about 0·8 μF per sq. cm. of surface. We shall refer to this again in a moment.

At 20° C. the time of discharge is less: presumably, therefore, the degree of discharge and the energy set free in discharge are less too. This is what we actually find: the amount of electricity discharged in the action current is smaller, and the initial heat is smaller, at the higher temperature.

Is 0·8 μF per sq. cm. of surface a reasonable figure for the capacity? A pair of parallel plates, distant d cm. apart, with material of specific induction capacity k between them, have a capacity per sq. cm. of $k/4\pi d$ electrostatic units or $1\cdot11 \times 10^{-6}\, k/4\pi d\ \mu\mathrm{F}$. Putting $k = 4$ as for oil we obtain $3\cdot5 \times 10^{-7}/d$. If this is equal to $0\cdot8$, d is $4\cdot4 \times 10^{-7}$ cm. This is about the length of the chain of a fatty acid molecule: so it would be possible to explain the observed initial heat by supposing that a unimolecular film on the surface of the nerve has a difference of potential of 50 mv. across it, which is partly discharged when the impulse goes by.

That this result is not altogether unreasonable is borne out by the fact that other methods show that certain living cells have surfaces which act as condensers with capacities of this order of size. It has been found, for example (Blinks, 1929 *a* and *b*), that when an E.M.F. is placed between electrodes situated on either side of the surface of a large plant cell (*Valonia* or *Nitella*), and the resulting current recorded with a string galvanometer, the times of charge and discharge correspond to a capacity per sq. cm. of $0\cdot1$ to $1\ \mu\mathrm{F}$ or more. By the use of high-frequency oscillating currents the surface of a red blood corpuscle (Fricke, 1925) has been shown to have a capacity of $0\cdot8\ \mu\mathrm{F}$ per sq. cm., which is independent of frequency down at any rate to 3600 per second. So far, therefore, as these calculations go, there is nothing impossible in the idea that the initial heat in nerve may be due to the partial discharge of an electrical double layer located at some cylindrical surface in the fibre.

In crab's nerve the initial heat per isolated impulse is much greater, at 20° C. 10 times as great, about $8\cdot5 \times 10^{-7}$ cal. per gram. The fibres are not much smaller: their surface, say,. is 3000 sq. cm. per gram. Repeating the calculation given above we should require a capacity of 2 μF per sq. cm. of fibre surface charged to 50 mv. and then half discharged, to explain the observed heat. This is stretching our credulity rather far. Crab's nerve may force us to discard the electrical explanation of the initial heat. Other considerations also (see Appendix 1) suggest a much lower value for the capacity, and it is not easy to believe in a potential gradient of 100,000 volts per cm., such as would correspond to 50 mv. in $4\cdot4 \times 10^{-7}$ cm.

THE PROPAGATED DISCHARGE
OF A POLARIZED FILM

It is assumed in this calculation that the film constituting the dielectric of the condenser remains "impermeable" or non-conducting so long as it is charged and at rest, but becomes "permeable" or conducting and permits neighbouring areas of the film to discharge through it when once it is itself discharged. A wave of discharge would then sweep along the transmitting surface. Let us see whether this agrees with known facts. Between an injured and an uninjured point in a nerve is a difference of potential which is generally reckoned to be about 50 mv. This is probably (fig. 2) a sign of a difference of potential across the surface of the nerve fibre: in the plant cells just referred to there certainly is such a difference of potential. The potential outside, therefore, is 50 mv. greater than that inside. Now all that is known of electric stimulation, and all theories of its mechanism, point to the fact that an electric current excites by changing the accumulation of ions at some membrane, or more probably by a sufficient discharge through the dielectric of the surface: the former is Nernst's well-known theory of electric excitation (Nernst, 1908): the latter is discussed below. The region of stimulation when a current is passed along a nerve is the cathode, the point where a negative current reaches the outside of the fibre. A stimulating current, therefore, does in fact tend to decrease, or to abolish, the difference of potential already supposed on very strong grounds to exist across the surface; and it does so most strongly at the point where excitation is known most readily to occur. The momentary depolarization of the membrane if intense enough causes an unstable state to be reached: it acts as a "stimulus" and an impulse starts off, each region discharging itself through the region already discharged, so itself becoming "stimulated", i.e. rendered permeable or conducting, and allowing the next region in turn to do the same.

It is a problem for physical and colloid chemists to devise a mechanism by which a sufficient discharge across a film or interface somehow allows that film for a moment to conduct, and so permits a neighbouring region of the film also to discharge through it. Not only, however, must it possess this property—it needs a much more formidable one: as soon as it is discharged, or partially discharged, it must be rapidly reformed ready to be discharged once more. The polar arrangement of the molecules is broken up perhaps on discharge and then reset: or an emulsion of water in oil becomes, for a moment, an emulsion of oil in water, and then rapidly reverts to its initial state.

THE ORIGIN OF THE RESTING ("INJURY") POTENTIAL

It has been supposed that the potential differences which occur at the surfaces of nerves and other cells are caused by differential diffusion. Beutner (1920) has devoted a monograph to the theory and it has been further developed by other authors (see e.g. Osterhout, 1931). It is known that potassium ions have certain peculiar properties. They are far more concentrated in the interior of muscle and nerve cells than in the fluid outside. They are almost certainly free in solution, for it is impossible to explain either the osmotic pressure, or the conductivity, without their presence as freely movable agents. (Their histological appearance at the surface under the action of precipitating substances is simply analogous to the formation of a copper ferrocyanide membrane in the pores of a Pfeffer pot.) Too high a concentration in the outside fluid (still far lower than inside) renders a muscle reversibly inexcitable (Dulière and Horton, 1929; Horton, 1930).

In crab's nerve, which is the simplest case, S. L. Cowan (unpublished) has recently found that by suitably varying the potassium-ion concentration in the bathing fluid the "injury" or resting potential can be made to take any value within wide limits.

On the diffusion theory the potassium ion is supposed to be able to pass through the film or interface bounding a cell, other ions being unable, or at least having a far lower mobility. If, for example, a KCl (inner) solution were completely separated from a NaCl (outer) solution by a membrane in which K ions could diffuse but Na and Cl ions could not, an electrical potential difference would be set up, the NaCl solution being positive.*

---

* A similar case is easily reproduced experimentally:

$N/10$ KCl | salicylic aldehyde containing salicylic acid | $N/1000$ KCl.
E.M.F. observed 0·1 volt.

If the membrane were injured at a point Cl ions could there diffuse out to meet K ions passing the membrane, or Na ions could go in to balance the K ions going out. The net effect would be that an "injury current" would be set up, the injured point being "negative". Current would pass in the "outer" solution from the uninjured to the injured point. This is just what happens in nerve.

In medullated nerve the matter is complicated by the existence of a double membrane, that on the outside and that on the inside of the sheath. It is impossible to make a simple picture of the result of such an arrangement. It is similar to that existing in the plant cells studied by Osterhout (1931, p. 382), which are surrounded by a layer of protoplasm covered with two films probably possessing different properties. In crab's nerve, which has no visible sheath, the conditions are simpler: probably here we are dealing with a single membrane. Recent work on this by Cowan in relation to potassium is so pertinent that, although not yet published, it is summarized (with his permission) here.

THE EFFECT OF POTASSIUM ON CRAB'S NERVE

Crab's blood contains on analysis about 0·05 p.c. potassium: crab's nerve about 10 times as much. The nerve necessarily retains blood between its fibres in spite of careful blotting, so the real concentration inside may be appreciably greater. Taking the ratio of 10 : 1 between inside and outside and supposing that K ions alone are able to penetrate the surface, the potential difference across it should be

$$RT \log_e c_2/c_1,$$

which is about 58 mv. Cowan has regularly observed values of 30 mv. in freshly dissected nerve and on two occasions of 40 mv., and since a certain amount of short-circuiting must occur in the strongly conducting fluid (about equivalent in composition to sea water) between the fibres, the real value may well approximate to the 58 mv. required by theory.

In the calculation it is assumed that K ions alone can penetrate the surface layer. A carefully dissected nerve ligatured at its ends and left at rest in oxygenated sea water allows no potassium at all to leak out, after the first hour or so, though it retains its full excitability. If K ions can traverse the surface, and we shall see in a moment from the effect of varying the K-ion concentration in the external fluid that they almost certainly can, then the fact that they remain inside in spite of a concentration ratio of 10 : 1 means that they are electrostatically held. If Cl, or any other anion, could come out with it, then the K ion would escape: if Na or any other cation could go in to take its place, then K would be able to go out. Therefore K is the only ion mobile in the surface layer, and the formula above is justified.

Osterhout has shown (1931) that the potential difference at the surface of a large plant cell may be altered over a wide range by varying the K-ion concentration in the external fluid. Cowan has found the same effect in crab's nerve. Soaking a nerve (which has no sheath round it, so mixing is rapid) for a few seconds in sea water to which K has been added, or in an artificial sea water containing no K, the potential difference may be varied from about 40 mv. to nearly nothing as desired. The effects, within this range, are reversible: if, however, the K-ion concentration be raised too high irreversible injury results.

This result is entirely in keeping with the view that the difference of K determines the resting potential difference.

Potassium solutions produce a further effect, in that they depress the conduction of the nervous impulse: with concentrations high enough the nerve becomes reversibly inexcitable.

When a nerve, which is leaking no K at rest, is stimulated to fatigue, K escapes during and for some minutes after stimulation. It seems, therefore, that the surface is rendered permeable during activity, not so much to K (to which on the present hypothesis it is always permeable) but to other ions (lactate?) the movement of which in the appropriate sense allows the K to diffuse in the direction of its concentration gradient. It may not indeed be a change of permeability at all, but simply the production of an anion capable of traversing the surface, and so of permitting the K ion also to escape.

These facts appear simple—there are complications, however, which prevent us from assuming that the surface of a crab's nerve is a simple physico-chemical membrane with fixed properties. For example (1) Ca ions delay the depression of conduction due to soaking in K solution: (2) lactate, bicarbonate and phosphate ions can certainly come out of a muscle cell (this has not been shown in nerve), so it is not obvious why K ions cannot. Perhaps the former come out in some undissociated form and dissociate outside: (3) the potential difference observed diminishes if the nerve be deprived of oxygen and rises again if oxygen be readmitted. As usual in biology, the simple physico-chemical explanation is inadequate. There can be no doubt, however, in spite of all complications, that the K ion and its diffusion through the surface are important factors in the development of the potentials observed in the conduction of the nervous impulse.

Cowan has recently found that rubidium and caesium ions in the external fluid have the same effect on the "injury" potential as K ions.

ANOTHER OBJECTION

There is another, and a rather subtle and conditional, objection to a simple electro-chemical theory of the origin of the potential differences manifested during activity: it takes no account of the "initial heat". It is true we have just calculated that this might be due to the discharge of a condenser, of capacity 0·8 μF per sq. cm. of nerve fibre surface, previously charged to 50 mv. or less. When, however, the condenser is recharged, if this were done by means of a diffusion potential of the type supposed, potassium diffusing out, for example, more rapidly than sodium diffusing in, the electrical energy would be derived entirely from the heat of the surroundings (in the case of the relatively dilute solutions involved in living cells) and the net thermal effect of each successive impulse would be *nil*. The diffusion potential theory, therefore, does not account for the initial heat: if we still regard the resting potential observed as due to the superior mobility of potassium in the surface film, and the transmitted disturbance as a momentary wave of depolarization or discharge, we have to invent some other process to explain the initial heat.

There is little reason why we should not, and the magnitude of the initial heat in crab's nerve, together with other considerations, rather suggests that we should. The electro-chemical theory of nerve transmission has in any case to suppose that the electrical discharge of any ring of the fibre surface somehow causes a change in it which lowers its electrical resistance and so allows the next ring in its turn to discharge: by this means the wave is transmitted. The sudden change of resistance reminds one a little of that which occurs in a neon lamp when it begins to discharge. We cannot avoid introducing this change of electrical resistance, this increase (as physiologists say) of permeability, this physical effect, whatever it is, of electric excitation. Moreover, the change is rapidly reversed: just as a muscle relaxes after it has contracted, so we have to suppose that the surface film of the nerve fibre rapidly regains its high resistance and so may be charged once more. There is much to say for the view that the nerve impulse propagates itself by electrical means, or as physiologists would put it, by the action current produced at each point stimulating the neighbouring point: we deceive ourselves, however, if we do not recognize behind this a cycle of molecular change in the nerve—which probably, in a world in which the entropy tends to increase, will require a provision of energy.

THE ANALOGY OF MUSCLE

The energy for muscular contraction is supplied in its initial, as distinguished from its recovery, phase by the hydrolysis of a compound of creatine and phosphoric acid, known by the name of "phosphagen". Immediately after contraction this substance is at least in part re-formed, the energy for its endothermic resynthesis being supplied by the breakdown of some carbohydrate compound to form lactic acid. This second phase occupies a few seconds after contraction and is followed by a third phase in which oxygen is used, and the energy liberated by oxidation employed in reforming the remainder of the phosphagen and in restoring the lactic acid to the substance from which it arose. The sequence of these three chemical phases is fairly certain, but we have as yet no idea of the mechanism by which the first of them is linked with the development of force and the doing of work.

There is unquestionably some extensive molecular change, during contraction, throughout the substance of a muscle—for example an alteration of double refraction occurs, of which photographic records have been made and found to run parallel to those of the mechanical response or of the production of heat (von Muralt, 1932). There is also a large change of viscosity during activity. One supposes that some molecular pattern, typical of the muscle at rest, gives place to another molecular pattern when it contracts. The molecular cycle is reversed when the muscle relaxes, just like the alteration imagined to take place at the surface of a nerve fibre: in muscle the change is slower and involves enormously more energy. It is not more difficult, however, to imagine a molecular change in a film, resulting in a rise and fall of electrical conductivity, than to picture a cycle in which the double refraction of muscle and its length and tension are reversibly altered.

AN ELECTRICAL BASIS OF NERVE
TRANSMISSION

I have several times assumed that the electric current produced at an active region is intimately associated with the mechanism of transmission of the nervous impulse: some would go further and say that the action current is the agent by which activity is propagated. This is suggested by a variety of facts:

(1) The impulse and its electric accompaniment travel with the same speed and have many common properties.

(2) The voltage and duration of the action current are such as will in fact excite: indeed the duration (see Appendix III) is about that of the most efficient stimulus.

(3) In an experiment described by Osterhout (1931, p. 389) transmission in the protoplasmic layer surrounding a large plant cell (*Nitella*) was blocked by injuring a region of the protoplasm: laying a salt bridge, to allow electrical conduction, between two uninjured points across the injured gap allowed activity to be transmitted. Here there was no possible continuity of a molecular layer between the two parts, and nothing that we know, except an electric current, could travel across the bridge.

ELECTRIC EXCITATION

The problem, therefore, of impulse transmission from point to point resolves itself into that of electric excitation—about which much has been written these many years. A nerve fibre is to be regarded as a cylindrical condenser,* with a source of electromotive force existing

---

* For a somewhat analogous, though not identical, discussion and for full historical references, see Ebbecke (1927).

in the dielectric between the plates. Normally at rest and unstimulated the outer plate is at a potential 50 mv. or so above the inner: this is the "resting" or "injury potential". When electrodes are placed upon the fibre and a current C is passed between them several things happen (fig. 10).

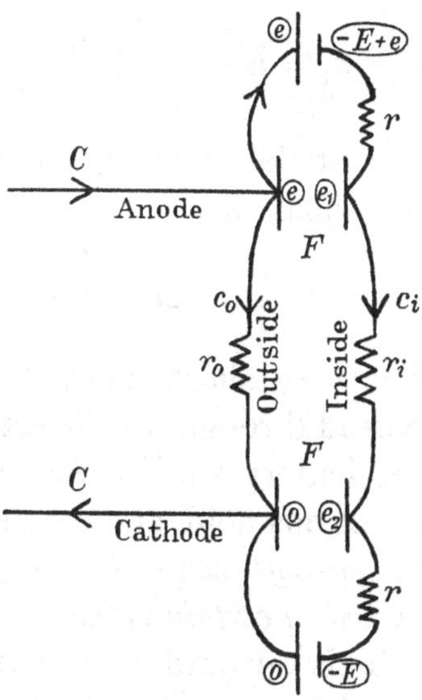

(A) Some—the chief part —of the current is short-circuited in the tissue fluids (current c_o through resistance r_o).

(B) Some of the current is absorbed by the electrical capacity of the fibre surface, the potential difference being increased at the anode, decreased at the cathode.

(C) Some of the current runs across the dielectric to the interior of the fibre, against the potential difference ("injury potential") E at the anode, with it at the cathode.

(D) This current completes its path by running down the interior of the fibre (current c_i through resistance r_i).

With the symbols shown in fig. 10 the potential differ-

Fig. 10. Electrical model of nerve stimulation. The current C is led in and out by the anode and cathode wires, the electrode regions of the nerve being represented by capacities F and F in parallel with sources of electromotive force E. The current C (i) charges the capacities, (ii) is short-circuited through the tissue fluids outside, and (iii) passes through the surface against the potential difference E and thence down the inside of the fibres.

The symbols given in rings represent electrical potentials at the points indicated: the c's are currents, the r's resistances. For calculations see text and Appendix I.

ence at the anode, initially E, rises according to the equation (see Appendix I)

$$e - e_1 = E + \frac{Crr_o}{2r + r_i + r_o}\left(1 - \epsilon^{-t \Big/ \frac{Fr(r_i + r_0)}{2r + r_i + r_0}}\right) \quad \text{......(i)},$$

that at the cathode, also initially E, falls according to the equation

$$- e_2 = E - \frac{Crr_o}{2r + r_i + r_o}\left(1 - \epsilon^{-t \Big/ \frac{Fr(r_i + r_0)}{2r + r_i + r_0}}\right) \text{......(ii)}.$$

For very short times, before the amount of current passed through the dielectric is large enough to polarize the battery and increase its potential difference, these equations hold, and we suppose *excitation to occur when the current outward through the dielectric at the cathode exceeds a certain value.*

This outward current (fig. 10) is $(e_2 + E)/r$ which is equal to

$$\frac{Cr_o}{2r + r_i + r_o}\left(1 - \epsilon^{-t \Big/ \frac{Fr(r_i + r_0)}{2r + r_i + r_0}}\right) \quad \text{......(iii)}.$$

Its initial value is zero—the whole of the current arriving at the electrodes (except the part short-circuited) is absorbed by the capacity at the surface. Its maximum value, apart from complicating factors, would be at $t = \infty$. Excitation, therefore, occurs at some time between zero and infinity, shorter the greater the current.

Much attention has been paid by physiologists to the experimental relation found in all excitable tissues between the strength of a constant current and the time it must be passed so as just to excite. This "strength-duration" relation is of the type shown in fig. 11. If we suppose that the minimum current, through the dielectric at the cathode, required for excitation is Q, then

$$Q = \frac{Cr_0}{2r + r_i + r_0}\left(1 - \epsilon^{-t \Big/ \frac{Fr(r_i + r_0)}{2r + r_i + r_0}}\right),$$

and the applied current is

$$C = \frac{Q(2r + r_i + r_0)/r_0}{1 - \epsilon^{-t \Big/ \frac{Fr(r_i + r_0)}{2r + r_i + r_0}}}.$$

A curve drawn from this equation is, at any rate approximately (indeed rather closely, see Appendix 1), of the type shown in fig. 11.

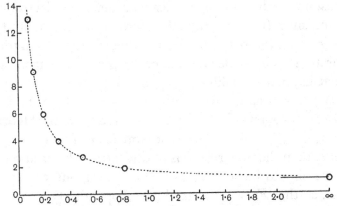

Fig. 11. "Strength-duration curve", constructed from Rushton's data (1932, p. 436, Frog iv) for frog's nerve, temperature "near zero". Horizontally the duration of the constant current used for excitation in σ, vertically the minimum strength necessary. The point on the right is for "infinite" duration and represents the "rheobase". The "chronaxie" or "excitation time"—i.e. the minimum time for excitation by a current twice the rheobase —is $0.72\,\sigma$.

It is known that the "strength-duration" curve alters in shape with various factors. It is interesting therefore to consider the quantity $Fr(r_i + r_0)/(2r + r_i + r_0)$ which determines the form of the curve.

(A) *Size of electrodes. Fr* is independent of the area of

the electrode region: if this be greater F will be greater and r proportionally less. When the size of the electrode region is made less, the distance between electrodes remaining the same, r becomes greater, Fr remaining constant, so the quantity $Fr (r_i + r_o)/(2r + r_i + r_o)$ diminishes. Consequently the "strength-duration" curve becomes more rapid. This is what happens in muscle (Lapicque, 1932; Watts, 1924; Jinnaka and Azuma, 1922). It is stated also (Grundfest, 1932) to be the case in nerve.

(B) *Distance between electrodes.* If Fr and r remain constant, a diminution of distance between electrodes diminishes $(r_i + r_o)$ and therefore also the quantity $Fr (r_i + r_o)/(2r + r_i + r_o)$. Consequently the "strength-duration" curve becomes more rapid. This again is what happens (Rushton, 1927, p. 361).

(C) *Absolute values.* With large electrodes very far apart $(r_i + r_o)$ would be large compared with $2r$. Hence $Fr (r_i + r_o)/(2r + r_i + r_o)$ approximates to Fr. In frog's nerve then the time relations of the "strength-duration" curve should be such that $\epsilon^{-t/Fr} = \frac{1}{2}$ at about 5×10^{-4} sec., so that $Fr = 3\cdot1 \times 10^{-4}$ (see Appendix 1).

(D) *Size of fibre.* If the diameter of the fibre increased, say n times, Fr would remain constant since F would increase and r diminish in proportion: r would diminish, becoming r/n, and r_i still more, becoming r_i/n^2. For simplicity, assume that r_o is small compared with the other resistances. Then the "excitation time" being proportional to $\dfrac{Fr (r_i + r_o)}{2r + r_i + r_o}$ would become

$$\frac{Fr \, r_i/n^2}{2r/n + r_i/n^2} = \frac{Fr \, r_i}{2rn + r_i}.$$

As n increases this diminishes, so that the time characteristic of a fibre of greater diameter should be less. This is well known to be the case. The velocity of propagation is greater and the excitation time is less, in fibres of greater diameter (see Gasser and Erlanger, 1927).

For an alternative explanation of some of these phenomena see Rushton (1932).

(E) *Effects of potassium and calcium.* Horton (1930) found a large increase in the excitation time of muscle by soaking it in a solution containing an excess of potassium: Lucas (1908) found a large decrease by similar treatment with an excess of calcium. The former means an increase, the latter a decrease, in the value of $Fr (r_i + r_o)/(2r + r_i + r_o)$. The only obvious way in which these ions could affect this quantity is by changing the value of r, the resistance of the sheath. It would be interesting to find, by Rushton's method (1927), whether the same ions similarly influence his "analytical unit of length": this should vary as $\sqrt{r}$.

"ACCOMMODATION"

The argument so far has not required any reference to the effect of currents lasting for more than a very short time. When a constant current is passed for a longer period the tissue becomes "accommodated" to it and no reaction is caused until it is stopped: then excitation occurs, this time at the anode. It is natural to regard this "accommodation" as an increase or decrease in the potential (E, fig. 10) developed across the dielectric when current is driven through it, rather like that in an accumulator through which a current is passed. The mechanism of this "polarization" of the battery we can easily see in the K hypothesis, if we suppose that the current in the dielectric can be carried only by K ions. At the anode the K ions in the fluid outside being in very low concentration are exhausted by their transfer through the dielectric, consequently the E.M.F. $= RT$ log (concentration inside)/(concentration outside) is increased. This resists the passage of the current till it stops. At the cathode K ions accumulate outside, and the ratio of the concentrations and consequently the E.M.F. also diminish. This "accommodation" we must picture as a relatively slow affair, and when a steady condition is finally attained no current will be running across the nerve surface and down its interior owing to the polarized state of the surface. The final condition therefore will be similar to that existing initially when undisturbed (except that E will be different at the two electrodes), so that the "break" of the constant current, being nothing more than the "make" of an equal constant current in the opposite direction, will have exactly the same effect as the "make" of the original current, but at the opposite pole. This is what is observed.

If this view of the "accommodation" of a tissue be correct it further explains why a slowly increasing current does not excite, even though it attains a value far greater than will excite if applied suddenly. With such a slowly increasing current the tissue is polarized, the counter E.M.F. is developed, before the current across the dielectric is large enough to excite.

We can see in this picture an explanation of the recent results of Erlanger and Blair (1931 *a* and *b*) who found that a sub-threshold current (impulsive or lasting) leaves behind it, in spite of its failure to excite, a condition of enhanced and then later of diminished excitability. The first phase is easy to understand: the second shock merely adds on to the first if they are close enough in time, the effect of the first not having been dissipated at once. In our model (fig. 10) the first shock partly spends itself in changing the charge on the condensers: if the second shock follows before these charges have normally reverted to their usual level (i.e. within a time not too long compared with that of half discharge, of the order of size of $Fr\,(r_o + r_i)/(2r + r_o + r_i)$) it is now able to complete the process started but not finished by the first. If, however, it comes later, the condensers have reverted to their normal state of charge, while the batteries are still somewhat polarized: in terms of the K theory the ions transferred by the current have not yet come back to their normal distribution, and they resist, more than normally, the passage of current across the surface which leads to excitation. In the phenomena described by Erlanger and Blair is another aspect of "accommodation": their results throw further light on it by showing how long is required for the polarization which underlies it to subside.

ALTERNATING CURRENTS

The effects of alternating currents of various frequencies can be explained on these lines.

(1) *Low frequency.* Here each phase provides a stimulus at the appropriate pole.

(2) *High frequency.* In the model (fig. 10), and according to equation (iii) above, the first result of a constant current is merely to change the charge on the

condensers, not to drive a current through the batteries; the immediate effect of each phase of an alternating current would be to do the same, and if the alternations were rapid enough *no current at all would go through the surface*. (The effect of high-frequency currents is apt to be attributed to the fact that such currents tend to run on the surface of conductors. With the relatively high specific resistance of living tissues this effect is negligible, except at frequencies far beyond the range at which the phenomenon is already very obvious.)

(3) *Medium frequency* (e.g. 1000 ∼ per sec.). If the nerve of a frog's muscle-nerve preparation be stimulated at such a frequency, after an initial response the muscle may completely relax. Stopping the stimulus and starting it again the same process is repeated. Failure occurs, not at the electrodes but in the nerve trunk or at the nerve ending. If we regard the agent in impulse transmission as electrical, the current by which any point "stimulates" the next puts it into a "relatively refractory" state. At the next stimulus the second point may not be excited at all, but according to Erlanger and Blair may nevertheless be put again into a refractory state: and so on. The basis of this explanation is that a stimulus may raise the threshold of excitation without actually exciting. This, on the present hypothesis, is due to the state of polarization in the surface left behind by the previous electrical disturbance.

THE CRISIS IN EXCITATION

Up to this stage the matter is fairly intelligible. The change, however, that happens when a sufficient current flows through the dielectric is not in the least understood. The electrical conductivity of the film at the surface suddenly rises when the current through it reaches a certain value; the potential difference in neighbouring regions is then short-circuited through the active portion: these in turn are discharged, their conductivity rises and a wave is propagated. The problem, above all others, which I want to submit to physical chemists, particularly to those with an interest in polarized surfaces, is—How can a rapid cycle of rise and fall of electrical conductivity follow an electrical discharge through a film? There are other problems, but this perhaps is the hardest and most fundamental. Its answer will go far to explain electric excitation, and therewith the mechanism of propagation of the nervous impulse.

THE IRON-WIRE MODEL

In recent years the views of physiologists about nervous action and transmission have been largely influenced by an electro-chemical model consisting of an iron wire in nitric acid. R. S. Lillie has made a variety of ingenious experiments with this; the phenomena encountered are often closely analogous to those observed in nerve. To quote Lillie (1923, p. 254), when an iron wire is immersed in a dish of nitric acid (about 60 p.c. by volume), after initial effervescence and darkening it rapidly reaches a "passive" condition, the surface becoming bright and covered with a thin film of oxide. If then it be touched, e.g. with zinc, or scratched, or "stimulated" with an electric current, a local reaction occurs, accompanied by effervescence and darkening, and sweeps rapidly over the wire. The local reaction ceases in 1 or 2 sec. and the metal reverts automatically to the passive state. It is then resistant to activation ("refractory"), but on standing the power of transmission returns.

The action of the nitric acid is twofold: it causes oxidation at the surface and the formation of an impermeable oxide film: and it acts as an acid, allowing the iron to go into solution as positively charged ions. If a current be passed through the solution into the wire and out again further on, the oxide film is chemically reduced at the point where the current enters the wire and a transmitted disturbance starts off. If the film is removed in any region the iron tends to go into solution there, and oxygen (as OH' ions) at neighbouring regions to complete the circuit. These regions, bared by the reduction of their protecting film, now in their turn send iron into solution and cause a reduction of the next area of film: the wave is propagated by the local "action current" rendering neighbouring areas active.

If the wire be enclosed in a tube filled with nitric acid the velocity of transmission is found to vary with the diameter of the tube: with a narrow tube the resistance through the electrolyte between neighbouring points is greater, the current is smaller, and the time required to reduce the film is increased. This recalls the way in which the velocity of transmission in a nerve fibre varies with its diameter. There are many such analogies, some of them pertinent, some of them probably not: the model is a good one provided that its limitations are realized.

The chief limitation is this: in the iron-wire model an impermeable film is formed over the whole surface, and the whole of this is chemically reduced as the wave goes by. The energy set free per sq. cm. of surface must be large: certainly not of the order of 10^{-3} erg as in nerve. The smallness indeed of the initial heat in nerve makes it quite clear that nothing analogous to the oxidation and reduction of a surface film occurs. If the wave of activity is transmitted by the electrical action of each spot on the next—as it very probably is—we must look for something more subtle, less drastic, than progressive reduction of a passive oxide film. Another difference—though a minor one—probably is that, whereas in the iron-wire model the current produced at the point A excites a neighbouring point B, in the nerve it is rather that when A is active B can discharge through A and so become active in its turn. With these limitations, and for those who prefer models to equations, the iron wire in its nitric acid can serve a useful purpose. It shows that such things as electro-chemical waves may be transmitted: it shows that their properties are connected and calculable: it is not, however, and it does not suggest, a theory of nerve activity.

THE REAL EXISTENCE OF THE INITIAL HEAT

Great stress has been laid on the initial heat production of nerve: on the heat which appears either during the passage of the impulse or so soon afterwards that no method at present available can separate impulse and heat. The problem of measuring this heat and placing it correctly in time is no easy one: the instruments have to be very sensitive and therefore rather slow. At 20° C. the time relations of nerve activity are so rapid that at first it seemed there might really be no initial heat at all. The instruments could not be made quicker: the nerve, however, could be made slower—by cooling to 0° C.

HWT

4

At that temperature a clear decision was reached—the initial heat existed, its maximum value per second was about $2 \cdot 2 \times 10^{-6}$ cal. per gram, per isolated impulse about $2 \cdot 6 \times 10^{-7}$ cal. per gram.*

ENERGY EXCHANGES IN MUSCLE

In muscle an analogous state of affairs exists: but here, owing to the much greater amount of heat available, the instruments need not be so sensitive, they can be made more rapid. When a muscle contracts and relaxes in response to a stimulus the heat is distributed in a certain characteristic way, quite independently of whether oxygen is present or not. During contraction there is an outburst of energy, probably derived from the hydrolysis of creatine-phosphoric acid ("phosphagen"): part of this energy appears at once as heat, part as mechanical tension. During relaxation such of this mechanical energy as was not used for doing work is degraded to heat. Together these heats (plus the mechanical work if any) are referred to as initial heat. Then a further process occurs, equally in the presence or the absence of oxygen: heat is set free, in a period of 20–120 sec. (at 17° C.) depending on conditions, in amount about 10 p.c. of the initial heat: this is the balance between two chemical processes, one endothermic, the other exothermic. Lactic acid appears, most of its energy being used to effect the resynthesis of creatine-phosphoric acid: a small remainder is wasted as heat. Finally, in—and only in—

* In crab's nerve at 20° C. the decision is even clearer, and according to recent work by Beresina and Feng the initial heat in a single impulse is about $8 \cdot 5 \times 10^{-7}$ cal. per gram, in 1 sec. of maximal stimulation about 70×10^{-6} cal. per gram. The recovery heat (lasting for 25 min.) is about 50 times as much (see fig. 8, p. 22).

the presence of oxygen, a recovery process occurs, some material is oxidized, and the energy so liberated is used to effect the resynthesis to carbohydrate of the lactic acid set free: part, however, of the energy is liberated as heat, in fact, the waste heat of recovery is about $1\frac{1}{5}$ times the initial heat. The recovery process takes many minutes: in the end, apart from a certain small amount of material oxidized, the initial condition is exactly restored.

We find therefore in muscle (see Hartree, 1932):

(*a*) Initial heat: relative value 1.

(*b*) Delayed non-oxidative heat: relative value 0·1.

(*c*) Delayed oxidative heat: relative value 1·2.

RECOVERY HEAT IN NERVE

It was natural to expect that nerve, which is analogous in many respects to muscle, would exhibit a similar distribution of heat. Two surprises awaited one.

In the first place there *is* delayed heat—but not 1·3 times the initial heat, about 15 times: this is for frog's nerve, given a few seconds' stimulus. In crab's limb nerve the ratio is greater, about 50 to 1. The conditions are clearly quite different from those in muscle: if the chemical reactions of the latter occur also in nerve in the same sequence, there is something else in addition, something of considerable importance.

In the second place, the whole of the delayed heat, not merely a fraction of it, occurs just the same whether oxygen is present or not. There is no doubt that ultimately the heat is derived from oxidation. This is shown by the facts (Gerard, 1927 *b*): (1) that in oxygen a nerve consumes extra oxygen when stimulated, in amount corresponding to the heat: (2) if kept and stimulated in nitrogen, it consumes extra oxygen afterwards when oxygen is introduced: (3) in nitrogen it produces CO_2 in measurable amounts. Apparently in nerve there is some oxygen compound, not dissociated even at a very low oxygen pressure, which provides a reserve for oxidation when molecular oxygen is excluded. The substance in question may really be of the nature of a hydrogen-acceptor, which is oxidized again when oxygen is admitted. Much research has been spent by biochemists on the hydrogen-acceptors which they find in crushed up organs. Here in nerve the physiologist can show them one actually at work in a living cell: I wish they would tell us how it functions.

THE "OXYGEN RESERVE"

There is no doubt of its existence. At 20° C. a frog's nerve remains excitable for about 2 hours when all oxygen is excluded: at 0° C. it may last for a day. During this time the response to stimulation gradually diminishes, but the distribution of heat in the several phases remains unaltered (Gerard, 1927 a; Hill, 1932). In muscle it is possible to separate the initial non-oxidative phase from the delayed oxygen-recovery one, simply by removing oxygen: in nerve apparently it is not. In the latter, as soon as recovery is made impossible by exhaustion of the oxygen—or potential oxygen—store, the whole process ceases. I imagine that recovery involves at least two things: (1) the restoration of a substance whose breakdown liberates the energy for the initial cycle, perhaps in the surface of the fibres: (2) the recharging of a battery of some kind by which the potential difference across that surface is maintained. If either fails, transmission must soon cease.

Here again is a curious contrast with muscle. Suspended in a suitable salt solution, entirely oxygen-free, at 20° C. a muscle, so thin that lactic acid formed in it can readily diffuse out, remains alive and responsive for days. Now muscle transmits just such an impulse as nerve: a wave of some kind, preceding the wave of contraction, passes along its fibres: why does the nerve and not the muscle fail as a consequence of oxygen lack? The answer is probably a chemical one. The muscle possesses in its lactic acid mechanism a device for carrying out all necessary restoration and recharging without any oxygen at all—so long as there is carbohydrate left and so long as the lactic acid can escape. Nerve apparently relies upon another mechanism—an oxygen, or a potential oxygen, reserve.

ALL CHEMICAL CHANGES MAY OCCUR IN RECOVERY

In the discussion of muscle we are inclined, without thinking about the matter much, to attribute the liberation of energy during contraction to some chemical reaction: this reaction used to be the formation of lactic acid: it is now the hydrolysis of creatine-phosphoric acid. A. D. Ritchie (1932) has pointed out that this is unnecessary, that we may perfectly well assume that the energy for contraction is stored in some latent physical form, requiring only to be released by a stimulus, the ensuing chemical reactions being successive stages in the restoration of this store of potential (e.g. electrical) energy to its initial state. His evidence is indirect, and it is difficult with present methods to see any means of obtaining direct evidence: but it is true that there is nothing pointing the other way. In nerve an initial breakdown occurs by which the impulse is propagated. During this, or immediately after it (and no instruments at present imaginable could decide which), there is an evolution of heat. It may well be the case that this so-called initial heat represents the first stage in recovery—corresponding, on Ritchie's hypothesis, to the phosphagen breakdown in muscle.

The delayed heat production in nerve, as in muscle, occurs in two phases. The first starts at a high rate and declines rapidly: its total amount is small, probably about equal to the initial heat: it is over in 20 or 30 sec. The second starts at a low rate but declines very slowly: its total amount is large: about 15 times the initial heat (see fig. 7). With the lessons of muscle behind us it is natural to regard these two phases as representing separate processes.* We suppose that a single impulse in a nerve is accompanied, or rapidly followed, by heat H during its transmission (H is the initial heat) followed by a rate of heat production ($Ae^{-at} + Be^{-bt}$) running to completion afterwards. The first, the A term, is the more rapid, a is greater than b: the second, the B term, is the greater, B/b is greater than A/a. By such equations the observed heat rate, during and after a stimulus of any duration, can be fairly completely described.

---

* In crab's nerve there seems to be only one—the slow—delayed process (see fig. 8). Feng (personal communication) suggests that the rapid fatiguability of crab's nerve depends on the absence of the first—the rapid—phase of recovery.

THE NATURE OF OXIDATIVE RECOVERY

Described but not explained. What are the two terms, and why is so much energy (relatively) spent in recovery? Let me suggest an answer by asking another question—Why does any living cell continually spend energy, even when it is doing apparently nothing at all? A resting nerve at 20° C. gives out by oxidation about 6.4×10^{-5} cal. per gram per second (Beresina, 1932): stimulated continuously at a maximum rate, its heat production is about doubled (Hill, 1932). Thus, doing nothing at all, merely existing in a state of readiness to respond, it is using about half as much energy as when giving its greatest response. There has been much speculation about this question: details are not certain, but in general it seems probable that a living cell is an unstable dynamical system which can be maintained only with the continual doing of work. For example, concentration differences tend by diffusion to disappear, potential differences to leak away: they must be kept up by some active effort on the part of the cell, energy must be liberated.

This is not mere speculation. The resting electrical potential of nerve is diminished, in crab's nerve it may indeed be abolished, by stimulation and by oxygen want: it is restored, or partly restored, if the nerve be left resting in oxygen. In a very different organ, the skin of the frog, normally, in the presence of oxygen, a difference of electrical potential exists between the two sides; if oxygen be removed, the potential diminishes: if oxygen be readmitted, the potential rises again (Lund, 1928). It is difficult to imagine that in such a system as a living tissue, potential differences can be due directly to any other cause than ionic concentration differences. When currents flow, ions must move. Now, according to thermodynamics, if the processes involved were reversible not much energy would be needed to separate ions which had mixed, to restore potential differences across films which had run down by long standing. Living cells, however, are like chemical manufacturers—they find it very difficult to come within sight of thermodynamic reversibility. The kidney, for example, which spends its life doing osmotic work—and hard work at that—has an average efficiency of the order of 1 p.c.: 99 p.c. of the energy of the materials which it oxidizes appears as heat, 1 p.c. only as osmotic work (Borsook and Winegarden, 1931). This may sound rather bad compared with a reversible cycle, but in addressing myself to chemists I will stand up for the living cell: I wonder whether chemical engineers would separate urine, or even $N/10$ HCl, from blood with an efficiency of one in a million, and if they did whether the blood in the end would be recognizable as such! The inefficiency of any actual process may well be the reason why so much energy is wasted—or apparently wasted—in resting metabolism, so much in the recovery process after activity in nerve.

The second phase, therefore, of recovery, my Be^{-bt}, where b is small and B/b is large, I tend to regard as the waste heat of an inefficient secretory process which forces back to their place ions which have escaped during activity, ions whose diffusion has caused the electric currents which we observe. It is no good as yet asking how this secretion is done: nobody knows, and nobody is likely to know for some little time: the models one reads of give very little assurance. That something like secretion, or recharging, occurs is fairly clearly shown when the resting potential of nerve, or of frog's skin, diminished by stimulation or by oxygen want, is raised again merely by a sojourn in oxygen.

What the first phase of recovery, my Ae^{-at}, is due to it is impossible at present to suggest. Its absence in crab's nerve may be significant. Were it not that Feng (1932) has shown that iodo-acetic acid (which stops lactic acid formation by affecting the catalyst of that reaction) does not abolish it, I should have attributed it to delayed lactic acid formation like the early delayed heat in muscle (Hartree, 1932). It is fortunate to have colleagues whose experiments anticipate one's errors!

So I will leave the matter—rather exciting, filled with pertinent possibilities, waiting for much more research. If only chemists, physicists and engineers would join in the investigation, and not imagine that the problem is either too easy or too hard! It is not too easy, as they would soon find out if they tried. Intellectually the problem is quite a respectable one. It is not too hard: there is no reason to think of it as insoluble, and it begins at once to yield to accurate measurement. The question outlined here needs a generation of workers as able, and of work as concentrated, as those that have been attracted to that of atomic structure. Then, instead of scattered phenomena and inconsistent deductions, such as I have presented, we might hope to possess a manageable theory of one of the most important facts in nature.

APPENDIX I

THE EXCITATION OF NERVE

The nerve (fig. 10, p. 39 above) is regarded as a cylindrical condenser (see also Ebbecke, 1927), a difference of potential being
maintained between the two plates by some source of electromotive force in the dielectric between them. A current C is
led in and out at two electrodes, the nerve surface at each being
represented by a condenser of capacity F in parallel with a
battery of E.M.F. E and of internal resistance r. It will be seen
below that the area of the electrode region is not of primary
concern, since the product Fr comes into the equations and this
is independent of the area of the condenser surface considered.

The condenser plates are merely the surface of the dielectric
and the fluid in its neighbourhood. Electronic conduction
along them is impossible. We will consider only the region
in which their charge is altered by electricity forced through
the dielectric by the stimulating current, i.e. near the electrodes.

Of the current C a large part, c_o, is short-circuited in the
fluids around the nerve fibre. When C is "made" its first effect
is to change the charge on the condensers at anode and cathode;
soon, however, as these charge up the current is diverted into
the other path shown in the figure: it runs through the dielectric, against the resistance r and the potential difference E
of the "condenser battery", and down the interior of the fibre.
It is supposed that excitation occurs when the current through
the dielectric exceeds a certain value.

In fig. 10 the symbols in circles show the potentials at the
points indicated: c_i and r_i are respectively the current in and
the resistance of the inside of the fibre between the electrodes.

It is required to find the current through the dielectric at
the anode and the cathode.

From Ohm's law
$$c_o r_o = e \qquad \qquad \text{......(i)},$$
$$c_i r_i = e_1 - e_2 \qquad \qquad \text{......(ii)}.$$

Considering the accumulation of charge at the outer plates of the condensers at anode and cathode respectively, we find

$$C = c_o + \frac{-E + e - e_1}{r} + F \frac{d}{dt}(e - e_1) \quad \ldots\ldots\text{(iii)},$$

$$C = c_o + \frac{E + e_2}{r} + F \frac{de_2}{dt} \quad \ldots\ldots\text{(iv)}.$$

Considering the rates of accumulation of charge at the inner plates,

$$c_i = F \frac{d}{dt}(e - e_1) + \frac{-E + e - e_1}{r} \quad \ldots\ldots\text{(v)},$$

$$c_i = F \frac{de_2}{dt} + \frac{e_2 + E}{r} \quad \ldots\ldots\text{(vi)}.$$

Subtracting (vi) from (v) and integrating, for the initial conditions
$$e = 0, \quad e_1 = -E, \quad e_2 = -E,$$
we find
$$e - e_1 - e_2 = 2E \quad \ldots\ldots\text{(vii)}.$$

Adding (iii) and (iv) and multiplying r_o, adding (v) and (vi) and multiplying by r_i, adding the results, and substituting for c_o and c_i from (i) and (ii), we find*

$$2Cr_o = (e - e_1 + e_2)\left(\frac{2r + r_o + r_1}{r}\right) + F(r_o + r_i)\frac{d}{dt}(e - e_1 + e_2).$$

Integrating for the initial conditions named above, we obtain

$$e - e_1 + e_2 = \frac{2Crr_o}{2r + r_i + r_o}\left(1 - \epsilon^{-t \big/ \frac{Fr(r_o + r_i)}{2r + r_o + r_i}}\right)\ldots\text{(viii)}.$$

Subtracting (viii) from (vii),

$$-e_2 = E - \frac{Crr_o}{2r + r_i + r_o}\left(1 - \epsilon^{-t \big/ \frac{Fr(r_o + r_i)}{2r + r_o + r_i}}\right) \ldots\text{(ix)},$$

which gives the potential difference at the cathode at time t. Similarly, for the potential difference at the anode,

$$e - e_1 = E + \frac{Crr_o}{2r + r_i + r_o}\left(1 - \epsilon^{-t \big/ \frac{Fr(r_o + r_i)}{2r + r_o + r_i}}\right) \ldots\text{(x)}.$$

* If it be required to know the individual values of e, e_1 and e_2, a further equation connecting them is obtained from (i), (ii), (iv) and (vi), viz. $C = e/r_o + (e_1 - e_2)/r_i$.

The rate of discharge of current through the "condenser battery" at the cathode is $(e_2 + E)/r$. This is

$$\frac{Cr_0}{2r + r_i + r_0}\left(1 - \epsilon^{-t\left/\frac{Fr(r_0 + r_i)}{2r + r_0 + r_i}\right.}\right) \qquad \ldots\ldots\text{(xi)}.$$

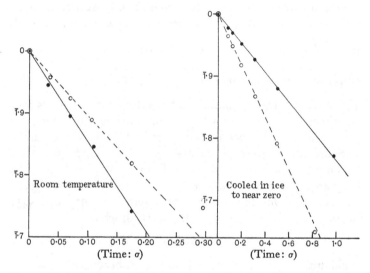

Fig. 12. To test equation (xii) in the text, relating the minimum strength of a constant current necessary for excitation to its duration. The equation being written,

$$C = \frac{R}{1 - \epsilon^{-kt}}, \quad \text{or } \log\left(1 - \frac{R}{C}\right) = -kt \log \epsilon,$$

$\log\left(1 - \frac{R}{C}\right)$ is plotted vertically against t horizontally. The experimental data (Rushton, 1932, p. 436) fit well on to straight lines. The slopes of these lines give the values of k.

When this is great enough excitation occurs. The initial rate of discharge is zero—all the current goes into the condenser. Apart from disturbing factors it rises to a maximum rate at $t = \infty$. If a constant current is to excite the maximum rate of discharge must reach a certain value Q say: the "rheobase", to use Lapicque's term, is

$$R = Q (2r + r_0 + r_i)/r_0.$$

For any duration t of a constant current the minimum current for excitation is

$$C = \frac{R}{\left(1 - \epsilon^{-t \Big/ \frac{Fr(r_o + r_i)}{2r + r_o + r_i}}\right)} \qquad \ldots\ldots\text{(xii)}.$$

This equation can readily be tested. The most accurate data available are probably those of Rushton (1932, p. 436) on frog's nerve. Expressing the above formula as $C = \frac{R}{1 - \epsilon^{-kt}}$, the simplest test is to plot $\log\left(1 - \frac{R}{C}\right)$ against t, when a straight line should result. In fig. 12 this has been done for Rushton's four experiments. Apart from a single point obviously in error there is no serious divergence from the straight lines. The equation therefore satisfies the experimental facts with sufficient accuracy.

The values of k can be deduced from the slopes of the lines. They are: (i) for the results at "room temperature" $k = 3360$ and 2380; (ii) for the results "near zero" $k = 955$ and 545. For the case of electrodes very far apart, in which

$$(r_o + r_i)/(2r + r_o + r_i)$$

would approximate to unity, equation (xii) becomes

$$C = \frac{R}{1 - \epsilon^{-t/Fr}},$$

so that $k = 1/Fr$.

Let us take F as the capacity, r as the resistance outwards, of 1 sq. cm. of nerve fibre surface. The only data available as to the values of these quantities can be derived from Rushton's calculations (1927). According to these the "analytical unit" of length, in which the resistance of the core of a nerve fibre parallel to the axis is equal to the resistance of its sheath perpendicular to the axis, is about 6 mm. The specific resistance of muscle at 20° C. (Hartree and Hill, 1921) is about 150 ohm-cm. Let us assume the same value for the core of a nerve fibre. A cylinder therefore 10μ in diameter and 0·6 cm. in length would have a resistance of $1\cdot15 \times 10^8$ ohms. The surface of such a cylinder would be $1\cdot9 \times 10^{-3}$ sq. cm. If the resistance of this surface were $1 \cdot 15 \times 10^8$ ohms, that of 1 sq. cm. would be $2 \cdot 2 \times 10^5$ ohms.

If $1/Fr$ were 2900 (the mean of the values at "room temperature" above) and r were $2 \cdot 2 \times 10^5$ ohms per sq. cm., F would be $1 \cdot 6 \times 10^{-9}$ farad $= 1 \cdot 6 \times 10^{-3}$ μF per sq. cm. This, with a value of 4 for the specific inductive capacity, would correspond to a thickness of dielectric $2 \cdot 2 \times 10^{-4}$ cm. $= 2 \cdot 2\mu$. It may be pure chance, but this is just about the thickness (see fig. 1) of the sheath of the nerve fibre.*

If 1 sq. cm. of sheath $2 \cdot 2\mu$ thick has a resistance of $2 \cdot 2 \times 10^5$ ohms, the specific resistance is 10^9 ohms, twice that given in Kaye and Laby's Tables for "conducting" glass, half that for gutta-percha. If the basis of Rushton's calculation is not affected by the fact that the surface has a potential difference across it and a capacity distributed along it, it is clear that nerve sheath is a material of rather high resistance.

---

* The formula for a parallel plate condenser has been applied as on p. 28 above. It would be more correct to use that for a cylindrical condenser: the accuracy of the data, however, does not justify the extra complication.

APPENDIX II

THE MEASUREMENT OF THE HEAT PRODUCTION OF NERVE

The methods used in measuring the heat production of nerve are described, in their latest form, by Hill (1932). The nerves are laid on the "hot" junctions of a thermopile with 150 very thin couples of constantan-iron insulated with bakelite: this is connected to a sensitive galvanometer. When the nerves are stimulated (by condenser discharges and a commutator) at a

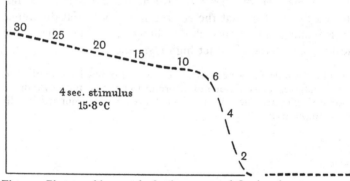

Fig. 13. Photographic record of galvanometer deflection due to 4 seconds stimulation of crab's limb nerves lying on the thermopile referred to in the text. Gaps in curve are time signals, numbered in seconds. The stimulus was immediately preceded by a rather longer gap (Beresina and Feng).

point several centimetres away from the thermopile, the galvanometer gives a deflection similar to that of fig. 13. This is recorded on photographic paper and analysed numerically by means of a "control" curve made by electrically heating the nerves from end to end with a known amount of energy.

It is necessary that the energy of the stimulus itself (see Appendix III) should not reach the thermopile: the nerves, between electrodes and thermopile, pass over a heavy block of silver which carries away the heat of the stimulus. "Stimulating" dead nerves produces no deflection.

The nerves on the thermopile lose heat rather rapidly, partly to the surroundings, partly to the "cold" junctions. The rapid upstroke is mainly due to the "initial heat": if there were no recovery heat the record would bend over and come back to the base-line. Owing, however, to the recovery heat it continues to move out at a slower rate, a maximum being reached (with crabs' nerves and with the thermopile used) in 2–4 min. (See fig. 8, p. 22, for the results of analysing a similar record.)

APPENDIX III

THE ENERGY OF A NERVE STIMULUS

The stimulating efficacy of a short constant current is a function of two variables, its strength and its duration. Expressing the minimum current C required to excite in the form given in Appendix I,

$$C = \frac{R}{1 - \epsilon^{-kt}},$$

its energy, for given electrodes, is proportional to $C^2 t$, i.e. to

$$\frac{t}{(1 - \epsilon^{-kt})^2}.$$

This quantity has a minimum value which can be found by differentiating: for minimum energy $kt = 1\cdot25$ approximately. For frog's nerve at room temperature k is about 3000 (Appendix I). Thus the duration t for minimum energy is about 4×10^{-4} sec.

It is much more convenient in practice to employ condenser discharges as stimuli: here again the stimulating efficacy is a function of two variables, the magnitude and the duration of the discharge. The energy of a condenser of $F \mu F$ capacity charged to potential V volts is $5 FV^2$ ergs. The duration of the discharge (to any ·given degree, e.g. $\frac{1}{2}$) is proportional to FR, where R is the total resistance through which discharge takes place. Again there is a "most efficient" or "optimal" stimulus, one by which stimulation is obtained with a minimum of energy.

Optimal values of FR have been found for frog's nerve by measuring the heat production caused by stimuli of constant energy but variable F (Hill, 1932, p. 125). At 20° C. the optimal value of FR is about 160, which corresponds to a time of half discharge of $1\cdot1 \times 10^{-4}$ sec. At 0° C. these quantities are 900 and $5\cdot8 \times 10^{-4}$ sec. respectively.

The efficacy of a stimulus may be varied in another way, viz. by adjusting the distance between the electrodes. With frogs' nerves the optimal distance is probably a few millimetres (Hill, 1932, p. 122).

With every adjustment made to get the most efficient stimulus the energy in it is still astonishingly large compared with that set free by the nerve itself when an impulse runs along it. For example, in a frog's sciatic nerve, in order to obtain anything like a maximal response the energy in a single stimulus must be about 0·25 erg. The initial heat in a single isolated impulse at 20° C. is about 7×10^{-8} cal., or about 3 ergs, per gram. A frog's sciatic weighs (say) 40 mg., so that the initial heat in the whole of it is about 0·12 erg. This is only one-half of the energy just calculated for a nearly maximal stimulus *at one point only*.

It is clear, therefore, that artificial stimulation by electric currents, even with every adjustment made to obtain the most efficient conditions, is very wasteful compared with the natural "stimulation" from point to point by which an impulse is propagated. This is easily intelligible. The chief part of the energy of an artificial stimulus is wasted in the tissue fluids between the electrodes (current c_o, fig. 10) outside the active fibres: only the fraction of current which crosses the cathode region of the surface is effective as a stimulus. In natural stimulation, i.e. from point to point, there are no electrodes and there can be no short-circuiting in the tissue fluids: the stimulating current therefore (if it be a current) is far more efficiently used.

The inefficiency of artificial stimulation has sometimes led to the belief that impulses set up by it are not the same as those which normally occur during life. There are really no grounds for this. Impulses "naturally" started, either in the central nervous system or in peripheral sensory organs, seem to have exactly the same characteristics as those which result a little distance away from an "artificial" shock. All the recent work of Adrian and his collaborators confirms this general statement. An impulse may be "artificial" within a few millimetres and a few units of 10^{-4} sec. from its point and time of origin: it is impossible to say, because of the leak of the stimulus itself into the recording apparatus. After propagation, however, by "natural" means over quite a short distance the impulse settles down to an ordinary transmitted disturbance.

It is unlikely, indeed, that an "all-or-none" process which

cannot be varied at all by varying the strength of an artificial stimulus, should be able to propagate itself in one form when started by "natural" means, and only in a different form when started by an electric current. If we find out the mechanism by which an impulse, started "artificially", is propagated we shall be close enough to a solution of the whole problem.

An attempt has been made by Winterstein (1931 *a* and *b*) to show that the increased oxygen consumption caused by stimulation of nerve is an artificial local effect at the electrodes and not due to impulses propagated to a distance. Apart from any other criticism of his views (Meyerhof, Gerard, Schmidt) it is quite certain that the heat production observed is due to a propagated and not to a local effect: heat locally produced at, or near, the electrodes, in the instruments employed, does not reach the thermopile. It would indeed be a remarkable fact if heat liberated by the propagation of impulses in a region several centimetres distant from the stimulating electrodes (many thousand times the diameter of the fibres involved) should be associated with no increase in the consumption of oxygen— particularly when no other chemical reaction to account for the heat has been found.

REFERENCES

AMBERSON, W. R. (1930). The effect of temperature upon the absolute refractory period in nerve. *J. Physiol.* **69**, 60.

BAYLISS, W. M. (1924). *Principles of General Physiology*. 4th ed. Longmans, Green and Co., London.

BERESINA, M. (1932). The resting heat production of nerve. *J. Physiol.* **76**, 170.

BERESINA, M. and FENG, T. P. The heat production of crustacean nerve. *J. Physiol.* **77**.

BEUTNER, R. (1920). *Die Entstehung elektrischer Ströme in lebenden Geweben*. Stuttgart.

BLINKS, L. R. (1929 *a*). The direct current resistance of *Valonia*. *J. Gen. Physiol.* **13**, 361.

BLINKS, W. R. (1929 *b*). The direct current resistance of *Nitella*. *Ibid.* **13**, 495.

BORELLI, J. A. (1710). *De motu animalium*. Peter Vander, Leiden.

BORSOOK, H. and WINEGARDEN, H. M. (1931). The energy cost of the excretion of urine. *Proc. Nat. Acad. Sci. Washington*, **17**, 13.

BROEMSER, P. (1929): Nervenleitungsgeschwindigkeit, Ermüdbarkeit, etc. Theorien der Nervenleitung. Article in *Handb. norm. path. Physiol.* **9**, 212. Springer, Berlin.

CREED, R. S., DENNY-BROWN, D., ECCLES, J. C., LIDDELL, E. G. T. and SHERRINGTON, C. S. (1932). *Reflex activity of the spinal cord.* Oxford University Press.

DULIÈRE, W. and HORTON, H. V. (1929). The reversible loss of excitability in isolated amphibian voluntary muscle. *J. Physiol.* **67**, 152.

EBBECKE, U. (1927). Über das Gesetz der elektrischen Reizung, etc. *Pflügers Arch.* **216**, 448.

ERLANGER, J. and BLAIR, E. A. (1931 *a*). The irritability changes in nerve in response to subthreshold induction shocks, and related phenomena including the relatively refractory phase. *Amer. J. Physiol.* **99**, 108.

—— —— (1931 *b*). The irritability changes in nerve in response to subthreshold constant currents, and related phenomena. *Ibid.* **99**, 129.

ERLANGER, J. and GASSER, H. S. (1924). The compound nature of the action current of nerve as disclosed by the cathode ray oscillograph. *Amer. J. Physiol.* **70**, 624.

FENG, T. P. (1932). *J. Physiol.* **76**, 477.

FRICKE, H. (1925). The electric capacity of suspensions with special reference to blood. *J. Gen. Physiol.* **9**, 137.

GASSER, H. S. (1928 *a*). The relation of the shape of the action potential of nerve to conduction velocity. *Amer. J. Physiol.* **84**, 699.

GASSER, H. S. (1928 b). The analysis of individual waves in the phrenic electroneurogram. *Ibid.* **85**, 569.

—— (1931). Nerve activity as modified by temperature change. *Ibid.* **97**, 254.

GASSER, H. S. and ERLANGER, J. (1927). The rôle played by the sizes of the constituent fibres of a nerve trunk in determining the form of its action potential wave. *Ibid.* **80**, 522.

GERARD, R. W. (1927 a). Studies on nerve metabolism. I. The influence of oxygen lack on heat production and action current. *J. Physiol.* **63**, 280.

—— (1927 b). Studies on nerve metabolism. II. Respiration in oxygen and nitrogen. *Amer. J. Physiol.* **82**, 381.

GERARD, R. W., HILL, A. V. and ZOTTERMAN, Y. (1927). The effect of frequency of stimulation on the heat production of nerve. *J. Physiol.* **63**, 130.

GRUNDFEST, H. (1932). Excitability of the single fibre nerve muscle complex. *Ibid.* **76**, 95.

HARDY, W. B. (1927). Molecular orientation in living matter. *J. Gen. Physiol.* **8**, 641.

HARTREE, W. (1932). The analysis of the delayed heat production of muscle. *J. Physiol.* **75**, 273.

HARTREE, W. and HILL, A. V. (1921). The specific electrical resistance of frog's muscle. *Biochem. J.* **15**, 379.

HILL, A. V. (1921). The energy involved in the electric change in muscle and nerve. *Proc. Roy. Soc.* B, **92**, 178.

—— (1932). A closer analysis of the heat production of nerve. *Ibid.* **111**, 106.

HORTON, H. V. (1930). The reversible loss of excitability in isolated amphibian voluntary muscle. *J. Physiol.* **70**, 389.

JINNAKA, S. and AZUMA, R. (1922). Electric current as a stimulus, with respect to its duration and strength. *Proc. Roy. Soc.* B, **94**, 49.

LAPICQUE, L. (1932). Retrograde polarization, a theory of systematic errors in measurements of muscular chronaxie through Ringer's fluid or with large electrodes. *J. Physiol.* **76**, 261.

LILLIE, R. S. (1923). *Protoplasmic action and nervous action.* Chicago University Press.

LUCAS, K. (1908). On the rate of development of the excitation process in muscle and nerve. *J. Physiol.* **37**, 459.

—— (1909). On the relation between the electric disturbance in muscle and the propagation of the excited state. *Ibid.* **39**, 207.

LUND, E. J. (1928). Relation between continuous bio-electric currents and cell respiration. III. Effects of concentration of oxygen on cell polarity in the frog's skin. *J. Exp. Zool.* **51**, 291.

MATTHEWS, B. H. C. (1929). Specific nerve impulses. *J. Physiol.* **67**, 169.

MONNIER, A. M. and DUBUISSON, M. (1931). Étude à l'oscillographe cathodique des nerfs pédieux de quelques Arthropodes. *Arch. int. Physiol.* **34**, 25.

MONNIER, A. M. and MONNIER, A. (1930). Étude du nerf de la ligne latérale de Mustelus Canis au moyen de l'oscillographe cathodique. *Ann. Physiol. et Physio-Chim. Biol.* **6**, 693.

VON MURALT, A. (1932). Über das Verhalten der Doppelbrechung des quergestreiften Muskels, während der Kontraktion. *Pflügers Arch.* **230**, 299.

NERNST, W. (1908). Zur Theorie des elektrischen Reizes. *Pflügers Arch.* **122**, 275.

NICOLAI, G. F. (1901). Über die Leitungsgeschwindigkeit im Reichnerven des Hechtes. *Pflügers Arch.* **85**, 65.

OSTERHOUT, W. J. V. (1931). Physiological studies of single plant cells. *Biol. Rev.* **6**, 369.

PARKER, G. H. (1932 *a*). Neuromuscular activities of the fishing filaments of *Physalia*. *J. Cell. Comp. Physiol.* **1**, 53.

—— (1932 *b*). On the trophic impulse so-called, its rate and nature. *Amer. Naturalist,* **66**, 147.

RITCHIE, A. D. (1932). Theories of muscular contraction. *Nature,* **129**, 165.

ROSENBERG, H. (1928). Die elektrischen Organe. Article in *Handb. d. norm. path. Physiol.* **8**, 2, 876. Springer, Berlin.

RUSHTON, W. A. H. (1927). The effect upon the threshold for nervous excitation of the length of nerve exposed, and the angle between current and nerve. *J. Physiol.* **63**, 357.

—— (1928). Excitation of bent nerve. *Ibid.* **65**, 173.

—— (1932). Identification of Lucas's α excitability. *Ibid.* **75**, 445.

WATTS, C. F. (1924). The effect of curari and denervation upon the electrical excitability of striated muscle. *Ibid.* **59**, 143.

WINTERSTEIN, H. (1931 *a*). Die elektrische Reizungsstoffwechsel der Nerven. *Biochem. Z.* **232**, 196.

—— (1931 *b*). Elektrische Reizung und physiologische Erregung. *Naturwissenschaften,* **19**, 247.

REFERENCES

INDEX